Best wishes,

Pat McGuire 9/6/07

towers to Fly !

Forty years after he started a college aviation program with no money
and no airplanes, look what's become of one cropduster's dream

FLIGHT OF THE ODEGARD

The inspiring, improbable saga of the takeoff and non-stop flight of the unparalleled
John D. Odegard School of Aerospace Sciences at the University of North Dakota

By Patrick A. McGuire

DEDICATED TO
THAT NICE IRISH KID
FROM LANGDON

Copyright 2007 by the UND Aerospace Press
Ryan Hall, Rm 211
4251 University Ave, Stop 9023
Grand Forks, North Dakota 58202-9023

Library of Congress Control Number: 2007929612

ISBN 978-0-9795975-0-3

Book design by Carol Roberts and Steve Larson,
FinePrint, Denver, CO

Edited by Richard H. Johnson and Bridey Orth

Prior page photo: The highly stylized, futuristic buildings on the campus of the John D. Odegard School of Aerospace Sciences are home to the nation's premier college of aviation education, atmospheric research, space studies, and computer science applications. With its 120 faculty members, 700 employees, 2,000 undergraduate and 200 graduate students from around the globe, it is today the second largest of the University of North Dakota's degree-granting colleges.

HIGH FLIGHT

Oh, I have slipped the surly bonds of earth
And danced the skies on laughter-silvered wings;
Sunward I've climbed, and joined the tumbling mirth
Of sun-split clouds – and done a hundred things
You have not dreamed of – wheeled and soared and swung
High in the sunlit silence. Hov'ring there,
I've chased the shouting wind along, and flung
My eager craft through footless halls of air.
Up, up the long, delirious burning blue
I've topped the windswept heights with easy grace
Where never lark, or even eagle flew.
And, while with silent, lifting mind I've trod
The high untresspassed sanctity of space,
Put out my hand, and touched the face of God.

by John Gillespie Magee, Jr

FOREWORD
His Ideas Were Always Big

By Senator Byron Dorgan

John Odegard and I first became friends in 1962 when we both worked summer jobs for the Boeing Company in Minot, North Dakota. If ever there was a round peg in a square hole it was John in his position that summer in the automated data processing department. Oh, he was good at his job. He just would rather have been flying! Imagine a rainbow forced to wear black and white.

U.S. Senator
Byron Dorgan

John was born to fly airplanes, but to do that he needed money. Three years out of high school, he had already found two stints in college not to his liking. So he worked whatever jobs he found to pay for flying time and aviation fuel. Still, he was not your standard college drop-out. He was a charismatic young man full of ideas. A person with boundless energy, he always seemed breathless about things, constantly pushing ideas forward and building momentum for what he wanted to do. And he always wanted to do much more than what he was doing.

Almost from day one of our friendship I felt that beneath all the ideas and energy lay real substance. I sensed that John Odegard was actually going to do those things he talked and dreamed of doing. I like to think I provided some insurance for that happening because it just so happened that Diane, his future wife, also worked at Boeing that summer. I introduced her to him and the rest—largely because of Diane's immovable faith in John—is the wonderful history laid out in these pages.

John and I kicked around together in those days and often went flying in an old single-engine Mooney. Inspired by his example, I learned to fly years later. But there was no comparing us as pilots. John Odegard wore an airplane the way you wear a suit. A man of disarming self-confidence on the ground, he took to the air so naturally and with such pure delight that I'm certain the air itself came to think of him as family.

Once, in the early eighties while we were flying into Grand Forks, one of our wheels wouldn't lock in the landing position. We flew around for awhile and had the tower check us as we went by. The tower said both wheels were down but in the cockpit a signal told us only one had locked. We really didn't know for sure. So John came in and landed on one wheel. After he had babied it to a stop, we saw that the locking pin indeed hadn't engaged in the suspect landing gear. It could have collapsed had we come in on both wheels. Through it all, John remained perfectly cool.

Flying takes a lot of patience and a lot of attention. John was always going a hundred miles an hour, his mind always moving. A dreamer, yes, but a realist as well. High in the air, he never lost sight of the ground.

We kept in touch over the years and when I got to a position in the United States Congress—first as a Representative and later as a Senator—where I could help John's aerospace programs at the University of North Dakota, I did that.

It never surprised me that John hatched the grandiose idea of creating an aviation school on the prairie; that he combined computer sciences—in which he became quite proficient—with aviation and then created the Center for Aerospace Sciences. I'm not sure John had a master plan for the many facets of the marvelous aerospace campus that graces the land and skies of Grand Forks today. I just knew that with John you kept peeling away the layers to find what the central idea was.

The central idea, of course, was to build. With his boundless energy John was a man very much in step with the wisdom of philosopher Carl Jung who said keep stacking rocks until a wall develops.

Lots of people come into my Senate office with various ideas and approaches. But no one ever quite came in with the flourish John had. His ideas were always big. He sat in my office in Washington one day and said "We've got to change the way people move in this country by air. We put all these people in the middle of a big city and 70 per cent of them are going elsewhere. Why not airports out in the middle of areas where there's no congestion?" He called them wayports, a whole new model of air transportation.

It was typical John, hatching a plan so very far above what you'd expect. Part of his technique was the "waterfalling" of ideas. Some were near and doable, others a little further upstream, building speed. If you were around him you got the impression that everything made a lot of sense and that his school was going to be successful and big. It was very easy to believe in John; he never gave you a reason to believe he wouldn't be successful.

That is why I bought in very early to the notion that he was building something extraordinary. I was able to persuade the administrators of the Federal Aviation Administration to come out to Grand Forks and see for themselves. Out of that came the FAA's designating The Center for Aerospace Sciences a Center of Excellence.

I did the same with Dan Golden, then the head of the National Aeronautic and

Space Administration. He was blown away by what he saw of the campus and at the end of his visit he said we need to do something out here. That's how the Odegard School and the University of North Dakota got involved with NASA's "Mission to Planet Earth," in which the massive body of data collected during space missions and from satellites is put to use to help people here on earth.

Since then I've worked to add more than $10 million in annual funding to the program. In the meantime NASA has loaned UND a DC-8 jet for various missions geared to collecting and analyzing space data that will aid programs here on earth.

The federal investment in The Odegard School is paying big dividends for the nation in other areas as well. I've been putting money into appropriations bills for years to support the Air Battle Captain program John started. It trains not only UND's R.O.T.C. cadets as helicopter pilots, but also a contingent each summer of 40 or more cadets from the United States Military Academy at West Point.

These freshly minted pilots go on to the Army's main helicopter program at Fort Rucker, Alabama and immediately qualify for the advanced course. That they have repeatedly finished in the top levels of their classes is just another way of demonstrating that in North Dakota we have one of the most respected flying schools in the world. The Odegard School has even been called "the Harvard of the air."

The kinds of technological innovation John envisioned and worked toward helped convince me in 2002 to initiate a research corridor program along the Red River Valley. I wanted to find a way to develop a high-tech economy in North Dakota so our kids could find jobs close to home. A recent report, prepared by North Dakota State University researcher Dr. Larry Leistritz suggests the program is doing just that. He writes that The Research Corridor Initiative has generated $759 million in positive economic impact and added thousands of jobs to the regional economy while growing those cutting-edge high tech industries in our state.

The Odegard School is one example of how such industries have grown in North Dakota. Our Red River Valley Research Corridor Initiative has fostered the start of the Center for Nanoscale Science and Engineering at NDSU, the Neurosciences Research Center at UND's School of Medicine, the National Center for Hydrogen Energy Technology at UND's Energy and Environmental Research Center, the National Energy Technology Training Center at Bismarck State College and the Center for Nanoscience Technology Training at NDSCS.

More recently we were able to draw directly on the expertise John helped establish at The Odegard School. We hosted a Red River Valley Research Corridor Action Summit in Grand Forks that brought together a variety of experts and policymakers to debate the future of Unmanned Aviation Systems.

The Odegard School and its dean, Bruce Smith, have taken the lead in helping support a new mission for the Grand Forks Air Force base—built around the use and development of unmanned aerial vehicles (UAV's). Indeed, North Dakota is increasingly becoming a hub of UAV activity through the efforts of the Grand Forks

Air Force Base, the 119th Fighter Squadron of the North Dakota Air National Guard—known by their nickname "The Happy Hooligans"—and the Center for Defense UAV Education at The Odegard School.

During most of his lifetime, John seldom received enough credit for his unique accomplishments. Where he saw great opportunity in the aviation industry and aerospace education, others resisted the change. But today he is properly viewed and hailed as a pioneer in these exciting, cutting-edge fields

While John never tired of promoting his ideas with a dynamic relentlessness you don't see very often, labeling him a promoter—which so often seems to mean self-promoter—doesn't do him justice. You would never see John call a press conference where you thought that he was promoting himself. The recognition he generated during his brief lifetime always connected to what he was building: aviation, computer science, the Center for Aerospace Sciences (renamed just before he died in 1998 The John D. Odegard School of Aerospace Sciences), his relationship with Northwest Airlines, cloud seeding in Morocco, the training of China Airlines pilots and Russian air traffic controllers.

John's persona became what his work was. That's as it should be, rather than a function of someone trying to become something for themselves. He really was the genuine article. I talked to him many times about politics and told him he'd be a natural to be part of North Dakota's leadership. He was accomplished, he was personally charming, he was smart. But John was always too busy to talk about those things.

Instead he continued reaching out to students, members of the local communities and national leaders. He was very persuasive and once he reached a critical mass of support, his school began moving on its own momentum.

The Odegard School is known today as one of the preeminent aviation schools in the world. When you're from a state which has the word north in its name and makes national news because it's 35 below zero, being known internationally for something that really matters, well, it makes one proud. For we North Dakotans see The Odegard School as something John built in our state.

We forget sometimes it was also his home state.

With his talent and energy he could have gone anywhere and commanded almost any job or salary. He chose instead to stay in North Dakota. For that and what he proved can be accomplished here, we owe him and his memory an unpayable debt.

Byron Dorgan is the senior United States Senator from North Dakota.

With its fleet of 120 single- and twin-engine prop planes, a half-dozen helicopters and two jets, The Odegard School operates one of the largest civilian training fleets in North America. The airport in small town Grand Forks may be tiny, clocking just six regular commercial flights a day to and from Minneapolis. But the constant daily takeoffs and landings make this one of the busiest airports in the nation.

CONTENTS

A lonely impulse of delight
Drove to this tumult in the clouds.
William Butler Yeats
An Irish Airman Foresees His Death

Prologue
An Impulse of Delight

It may be argued that the most dramatic moment in the history of anything is the teetering, fragile instant when something replaces nothing: when concrete hardens sand, when blurring motion eclipses talk, when trembling wings catch the lift of dreams.

Many are the stories of human achievement down the centuries that are told and retold in the form of reverse legerdemain—Now you don't see it, Now you do. The story of Flight? It's the dramatic moment when Orville Wright wobbled for 12 seconds over the sands of Kittyhawk. The dramatic story of the telephone? Why, there is Alexander Graham Bell shouting into a tube, "Watson, come here, I need you." The story of Rock and Roll? Hard not to hear Bill Haley singing, "One, two, three o'clock, four o'clock rock…"

For the most part, though, the grand renderings of history tend to emerge into consciousness without music, press release or Hollywood clarity. One has only to consider the Internet. Anyone of a certain age can easily recall a time when e-mail—an everyday, taken-for-granted necessity—didn't exist. But very likely, not one in ten thousand can recall the teetering, fragile instant when ordinary ether went from the stuff they gave you when they took out your tonsils to the metaphoric dark void wherein bloggers and web riders exchange invisible lasers of doggerel and brilliance.

Just because we don't remember the moment doesn't mean it didn't happen—although it's a good bet that if we don't remember it, then it wasn't particularly filmable. For in the deep space between a vision and the visible, one finds dozens, hundreds of scarcely remembered milestone moments that hardly qualify as dramatic, let alone remarkable. Without them, almost any given success might never have been achieved.

Thomas J. Clifford, for example, doesn't remember the weather or what he was wearing or even the precise date that it happened, but a subtle something that he did on a spring day in 1964 in his office as Dean of the Business School at the University of North Dakota—something he had done before and had just that day sworn not to do again—set in motion a series of events that would change the lives and fortunes of thousands of students, teachers and parents around the world.

It was Easter time, and in a conversation with Registrar Ruby McKenzie, Clifford had been discussing his notorious soft spot for admitting the occasional marginal student to the university. That he even had a soft spot would have surprised anyone reviewing the Clifford file in those days. He'd grown up a tame-resistant, cheerful farm boy in Langdon, North Dakota. The term in those days was hellion, a label used to defame a young devil. To the poetic soul of an untamed Irishman, however, it can also suggest a restless youth struggling against the one-size-is-supposed-to-fit-all standard boyhood model. Sabatini's *Scaramouch* set the standard: "He was born with a gift of laughter and a sense that the world was mad."

Clifford's father died when he was 12. In the emotional upheaval afterward, even the iron hand of the local Christian Brothers lost its grip on him. Wild became wilder. Regular fist fights escalated to the day when young Clifford shot a neighbor's cat from his bedroom window. It was left to his mother to salvage something of her rough and tumble son. She strongly believed in education and was so determined that he go to college that she packed up young Tom and transported him down to Grand Forks herself.

While at the university he played football and, with the onset of World War II, enlisted in the U.S. Marine Corps. In the Pacific, in horrific places such as Guadalcanal, Iwo Jima and Saipan, Clifford's wildness was tempered into practical steel. Flip through a yearbook from his training days as a marine officer candidate, and you will see page after page of earnest young men who never came back from the war.

Clifford himself was wounded three times, once blown completely out of the tank he commanded. His heroism under fire earned him silver and bronze stars and three purple hearts. He was best remembered by those who served under him as a rock-hard leader with a sense of humor; a man who would go to great lengths to keep his men alive. Decisive, though always a bit mischievous, he commanded rather than demanded the loyalty of his troops.

When he returned to Grand Forks after the war, those who thought the hammer of battle might have pounded the wild man into submission were proven wrong. The war had not so much changed Clifford as it had intensified his persona; he came back a toughened, supremely confident young man, modified incongruously by an alluring, seductive charm. Still carrying un-excised shrapnel from his war wounds, he'd convinced himself he would die young. It explains, perhaps, his knack for seeing life as a dance of comedy and tragedy: something to be enjoyed but to be wary of at the same time.

An accounting graduate before the war, Clifford qualified as a C.P.A. on his return and went on to earn a law degree at UND. He became a favorite of President John West, who named him the university's youngest ever Dean of Students. He later earned an M.B.A. at Stanford and, under UND President George Starcher, became the youngest ever Dean of the School of Business. Later, Starcher gave him the additional title of Vice President of Finance.

Clifford approached his duties with an almost fatalistic aggression. With his engaging personality and razor wit, he became a favorite of politicians who found it hard to say no to him. He used such capital to fuel a bold, risk-taking leadership style seldom seen on the university campus. He scorned the caution of academe, skating the margins of tradition and trouble to the tune of a miscreant's rationalism: it's easier, he would say, to ask forgiveness than permission. In other words, get it done and worry about the consequences later. Certainly he was unfazed by any of the traditional taboos or bureaucratic dictums that normally keep university faculty in a knot. When you've survived combat on Iwo Jima, he would say, nothing else worries you ever again.

Though he was a man who knew what he wanted and was not to be crossed, Clifford was a soft touch from the start with students who'd come out of the service. More than once he refused to excommunicate a tipsy veteran who'd let off too much steam after a hard night of off-campus "study." He was known to slip a few dollars to needy married students and even to help others with their homework. As his own man, Clifford found friends in both high and low places and even a few enemies in the ivory towers. They resented him because in spite of his war record, his C.P.A., his M.B.A., and his J.D., he was not an academic. In fact, Clifford never claimed to be an academic, reveling in his non-membership in that vaunted club.

But during that 1964 Eastertide meeting with Registrar Ruby McKenzie, Clifford conceded that perhaps he'd been a little too sympathetic to the needy, wayward student-type and not discriminating enough on behalf of the high standards of the university. He vowed to take a tougher stance, to finally draw the line against admitting students whose record showed they just weren't college material. His resolve would soon be tested.

Into the registrar's office that day came a brash 23-year-old named John Odegard, looking to transfer to UND from Minot State University. Among his credentials: he had never read a book his entire time at Minot. He had never done homework, nor had he ever studied. In fact, he had dropped out. The grades he left behind were mostly in the sub-basement of the alphabet.

One dubious plus: he had a pilot's license—although it once had been

Tom Clifford, then dean of the business school, admitted John Odegard to UND in 1964 in spite of his poor grades. Clifford saw a lot of his own wild youth in Odegard and became his mentor.

John Odegard and Diane Rosedale met in the summer of 1964. He was reckless, untamed and charming. She was refined, serious-minded and practical. She encouraged him to get his degree and to stop skydiving.

suspended by the Federal Aviation Administration. Even today there are multiple versions as to why. It was said by some that while flying over Minot, Odegard had thrown rolls of toilet paper from his plane and then had dived low to try to slice off the trailing streamers with his wings before the roll hit the ground. Supposedly, an FAA officer had seen him and copied down the number from his wings.

But no, say others. That was a different incident. The one he got suspended for was when he was flying his plane over a small lake, skipped the wheels over the water and then zoomed up and off a water ski jump. Once again, according to legend, a passing FAA officer saw him and jotted down the number on his wings.

Maybe so, say friends, but what about the time he was out flying and saw some buddies in a car down on the road below. To get their attention he dropped down and rolled one of the plane's wheels over the roof of their moving car. Didn't an FAA officer see him? Yes, perhaps that was it. One thing no one disputes, however: in addition to losing his pilot's license for awhile, Odegard also stirred up trouble in Minot when he showed up for his automobile driver's test—behind the wheel of his own car.

All of that was almost small change compared to the time he was suspended from Minot High School for rigging a chair with electrodes and then, as a gag, flash frying an unsuspecting classmate. It was that incident that prompted a counselor to warn Clara and Truman Odegard that their son, as popular as he was with his classmates, would never amount to anything—"not even a ditch digger"—because he was "completely irresponsible."

After reviewing young Odegard's record that spring day of 1964, Ruby McKenzie did what one hundred out of one hundred other registrars would have done. She shook her head and told Odegard no can do. Odegard, however, didn't take no for an answer—a trait that would later come to define him. With his fiancée Diane Rosedale in hand, he appealed to the next highest authority, the freshly hardened and former push-over Dean, Tom Clifford. A review of the record gave Clifford the same impression of Odegard as Ruby McKenzie had formed. But there was something else. As Odegard's flying indicated, he was reckless and untamed. Yet Clifford found him affable and supremely confident. Just the kind of young man who could stand some tempering. At the same time, Odegard seemed to be that familiar sort of male who needed a strong-willed woman to haul his ragged attitude off to the university.

Diane Rosedale was a refined, much more serious-minded sort than her fiancée. A soon-to-be graduate of Milwaukee-Downer College, she spoke fluent French. She and John were to be married in June, she told Clifford, emphasizing his need for a career. At her urging, John had decided to go back to school, finish his undergraduate degree at UND and then get an advanced degree in accounting. His goal: become a certified public accountant.

Clifford, of course, had already formed an opinion that the oxymoronic terms "reckless" and "C.P.A." were not necessarily oil and water. He certainly recognized the type of young man he was dealing with. The actor John Wayne, playing the role of Rooster Cogburn in the movie "True Grit," captured it perfectly. As Cogburn, describing the young, untamed Mattie Ross, Wayne had exclaimed, "By God, she reminds me of me."

As he had done so many times before, Clifford ignored the one-size-has-to-fit all model of the male persuasion. He went with his gut and gave Odegard his chance. Upon hearing that he had decided to admit this less than promising student, Ruby McKenzie was aghast. "But I thought you weren't going to…"

"Ruby," Clifford shrugged, indicating Odegard and his fiancée, "I changed my mind. Anyone smart enough to be with her deserves to get in."

Only in the rear view mirror of time does that story ooze significance. At Easter time, 1964, the John D. Odegard School of Aerospace Sciences didn't exist, not even as a wild idea in John Odegard's restless, churning mind. Today, because of Clifford's change of heart and the boundless energy of the man who became his protégé, there exists on a patch of windswept North Dakota prairie the nation's premier college of aviation education, atmospheric research, space studies and computer science applications. In fact, with its 120 faculty members, 700 employees, 2,000 undergraduate and 200 graduate students from around the globe, it is today the second largest of the University of North Dakota's degree-granting colleges. It is also the operator of one of the biggest civilian training fleets in North America.

It would have been impressive enough had such an institution risen from the mist in a storied setting like California or Texas or New York City. To have accomplished such an achievement in the blizzardy outer space of North Dakota, as Odegard did over a 30-year period, is almost to stretch the limits of credibility. Bob Muhs, one of Odegard's hundreds of loyal aerospace graduates, still shakes his head in wonder: "There's no reason on God's green Earth such a school should be in North Dakota."

Except, it is not only there, it thrives there. The airport in small town Grand Forks may be tiny, clocking just six regular commercial flights a day to and from Minneapolis. But the constant daily takeoffs and landings of 85 single- and twin-engine prop planes, along with a half-dozen helicopters all piloted by student fliers or cadets from the U.S. Military Academy at West Point, make this one of the busiest airports in the nation.

With apologies to Muhs, had an aviator been part of God's green Earth creation team, North Dakota's flat, see-forever panorama of wheat fields and ranch land—it's said if you stand on a tuna can you can see Seattle—could easily have served as the drawing board for the sky. Though winter months often register temperatures in the twenty to thirty below zero range, fliers know that aircraft perform better in the cold. The air is clear and more dense, and while it may make life difficult for humans, engines eat this stuff up.

The wind adds another valuable element to the North Dakota factor: when teaching students to fly, you don't always want it calm. Even during overcast conditions the ceilings are high enough to safely comply with Visual Flight Rules (VFR) requirements. In the meantime, a management information system warns schedulers when the ambient temperature gets colder than 25 degrees below zero or 35 below with wind chill—the practical limits for such equipment. Line maintenance crews put aircraft to bed at night in reverse order of the following morning's flight schedule. Students go out the next morning and conduct their preflight regimen in a heated hangar. They haul their plane out, fly their scheduled times and after landing, put a cowling cover on the engine to keep it warm for the next student crew. As a result, from January through December, through snow and blizzard, only two percent of all available flight time has to be cancelled because of weather.

Over the years, the Odegard School has trained pilots from Taiwan and mainland China, air traffic controllers from Russia and cloud seeders from Morocco. It recently signed a contract with Tokai University to train pilots for Japan's All Nippon Airways. Each summer, a corps of cadets from West Point comes to Grand Forks to learn how to fly helicopters. When they graduate and move on to Fort Rucker, the site of the Army's own helicopter training school, those North Dakota-trained West Pointers—along with UND ROTC cadets—consistently finish at or near the top of their class.

Learning to fly, however, is but a small piece of the Odegard mission. The school is also home to a Regional Weather Information Center, where its weather modification products help farmers become more in tune with their environment, and therefore more productive. Its researchers have developed computer algorithms for digital Doppler radar early warning systems to keep commercial pilots from flying into deadly micro-bursts. Every day, its Aerospace Network transmits class lectures, conferences and training materials to schools, industry sites and government facilities across the world for industry professionals looking to upgrade their skills. While many of the school's graduates have become airline pilots, many more have become executives for airline companies and for scores of firms and government agencies that deal on a daily basis with the ever-expanding industries of aviation and aerospace.

The school's gleaming $140 million complex of futuristic buildings and walkways have been welded incongruously onto the Western edge of the quaint, ivy covered, very traditional University of North Dakota. Two opposing cultures—the techno-modern and the traditional academic—have more or less blended over the decades, but not always smoothly and not without suspicion and jealousy.

Still, for those coming into Grand Forks for the first time, this sprawling aviation empire is often the first part of the university they see, from the five-story office/classroom complex at the airport to the 1,600 foot covered walkway that reaches across 42nd Street to link the futuristic Ryan and Clifford Halls. A guide will most likely mention two things at the sight of this walkway. The first is that not one dime of state or local money went into any of those buildings or airplanes or hangars or skywalks or programs. It came either from private donors and investors, business revenues, bonds or the federal government. Second, just one man got that money from out there to here; one man got those buildings built, those students hooked, those professors recruited and hired. One man, named John Odegard.

In fact, neither statement is completely true.

First, the money. In 1967, when the ambitious Odegard began talking about starting an aviation program at UND, he made the only bargain he could think of that would get him permission from the state and the university to go forward with a plan most people thought imbecilic at best. He promised he would never, *ever* ask the state or the university for one penny of funding. It would cost them nothing, he said, so what did they have to lose?

In the back of his mind, Odegard figured that once he showed some success then his promise would be forgotten and the state and university would gladly— *proudly*—fund him until the Red River of the North dried up. Being a native North Dakotan, he should have known better. For it is known in the land of the flickertail that if you promise a North Dakota legislator never, *ever* to ask for a penny, not only will that legislator remember said pledge in perpetuity, but his or her heirs will hold that promise sacred, as will their heirs and every heir from Fargo to Williston. And even should you lay starving on the side of the road begging for just one penny, that promise will be rolled up like a newspaper and rapped smartly against your sorry snout. Turnips will bleed before you see even a ha' penny.

In truth, an unkeepable promise of "Never!" often begets an equally untenable response of "Never!" On the theory that two negatives make a positive, it's not surprising that some state money has trickled down to the Aerospace coffers—a hundred thousand here and there in the early years—with said funds usually coming through a delicate priming of the political pump, or a reaching into some slush fund or other. Still, it is a fact that the campus buildings themselves, the most visible part of the school, were primarily funded by federal monies. Half of one of those buildings—Ryan Hall—and all of UND's airport structures were funded from other, non-public sources. Which brings us to the bent truth of the statement that John Odegard single-handedly got that money and created the aerospace school.

While Odegard's importance to the school's existence is quite obvious, the role of his mentor, Tom Clifford, is less so, yet just as critical. As a UND vice president, Clifford's tolerance and encouragement of the soon-to-explode wild dreams of John Odegard was invaluable. Not too many years after Odegard was admitted to the university, Clifford became its president. For the next 20 years Clifford demonstrated his belief and trust in Odegard by giving him room to develop what quickly became a lifelong project: start a school that would not only train students to fly but would give them an intense liberal arts education and enough business acumen to become a professional in the aviation industry, whether in the cockpit or in the boardroom.

Simultaneously, Clifford deflected endless protests from the disdainful faculty across campus. Outraged that a university should admit pilots into the lofty world of academe, they mounted seething campaigns aimed at undermining and getting rid of Odegard.

"Perhaps another university president would have shut it down," says Byron Dorgan, North Dakota's senior United States Senator. "There was controversy, there was difficulty," says Dorgan, a longtime friend of Odegard's from his early Minot days. "You've got this young whippersnapper who has an idea that will ruffle feathers. Most presidents would have said, 'I don't need this grief here. I don't even know if it will work.' Tom had a lot of guts and was willing to believe in John. And John felt very strongly that the support Clifford gave him was absolutely critical."

Clifford never had an interest in airplanes or aviation, but he saw a benefit to the university if things indeed developed the way the optimistic Odegard promised. He also admired the work ethic of the man who was supposedly too irresponsible to be a ditch digger. Odegard worked long and hard and always seemed able to make a dollar out of a dime, even if he often bent the rules and frequently stretched Clifford's patience. In the end, results meant everything to Clifford. While he had that soft spot for the needy student, he was still, after all, a marine. He always made it terrifyingly clear to those he empowered: here's your chance; make something of it or you're in the wind.

Clifford's charm was such that in his years at UND, he developed a far-reaching network of people eager to do something for him. His talent went beyond the mere befriending of and doing favors for powerful people across the state; he was a master at the delicate process of calling in a marker at exactly the right moment. With Odegard's reach often exceeding his grasp, Clifford was forced on many occasions to place a phone call to a key player across the state or country— sometimes in the public eye, sometimes in private industry—to turn up a hundred thousand or so to keep the planes aloft and his protégé out of the fire for another month or so.

It really comes down to this: Without Clifford, no Odegard. Without Odegard, no aerospace school. Without the school itself, the university never would have had colorful giants like Col. George Hammond, or Admiral Robert Shumaker or aviation pioneer Don Smith or astronaut Buzz Aldrin or dozens of other aces in their respective fields adding their considerable wisdom, prestige and influence to the program.

"With all due respect, John was a dreamer and visionary, but he didn't know how to get things done says Dana Siewert, today director of safety operations at the aerospace school and one of Odegard's original lieutenants from back in the day. "But John had the knack of surrounding himself with people he could trust and whom he knew could get job done."

Without such dedicated loyalists, it's likely the enrollment today at the University of North Dakota—the state's crown jewel—would sit closer to 10,000 than its current 13,000. Indeed, the crown might belong to someone else. As for the state itself—contentedly invisible on the national stage as a hearty enclave of self sufficiency, neighborly loyalty and family values—without the Odegard School, it would probably still be confused by outlanders with North Carolina or Montana. And to would-be tourists and second-tier geographers, it might still be best known, erroneously, as the home of Mount Rushmore. To those who believe their state is the best-kept secret in America and like it that way, this wouldn't be a problem. But to those trying to promote the state and to attract new business and to arrest the draining outflow of young people to the big cities, The Odegard School is a wondrous tool.

"I was in Taipei a few years ago, meeting with the president of Taiwan," recalls Byron Dorgan. "And he said to me, 'Your state trains pilots for our airline.' I said to my wife, 'That's so typical. You go halfway around the world and you hear about John Odegard's work from the president of a foreign country.'"

If one were to tell the story of a tree, it would be one thing to say, "Lo, a tiny sapling grew into a giant sequoia." Quite another to say, "A giant among redwoods has fallen; count its many rings, walk its length, feel its rough bark, see how its thick branches still reach out forever. And see the great dark hole it has left in our forest. Whatever shall we do?"

In the first version there is little to distinguish one giant sequoia from any other. In the second, it's the tragedy of the fall that brings this particular tree to our attention; examining what is left behind, we come to understand the enormity of the thing, the space it filled, the almighty struggle it faced to climb to the top of the forest where it could reach out to the sky.

Which is why, for all its delightful irony, Tom Clifford's change of heart in 1964 isn't really the most dramatic place to start this story.

The death of John Odegard in 1998 to cancer—34 years after the moment his fate and Clifford's and the University of North Dakota's were forever sealed— was indeed tragic. It could be said that his death was far more earth shaking than any of the improbable adventures Odegard starred in during those three-plus decades of leaping impossible obstacles and landing on both feet with hands full of gold. For all of the drama within those frenzied years—dramas to be played out in all their glory, sweat, passion and humor in the following pages—the most stunning moment of all seemed to come not at the beginning, or at the middle, but at the end. Churchill would have called it, more appropriately, the end of the beginning. For The Odegard School did not dry up and blow away with the passing of its king, dictator, prime-minister, despot, mentor, patriarch, CEO, chief pilot, strongman, cheerleader, father-figure, teacher, rainmaker, comedian, tyrant, wizard, and non-stop master of ceremonies. To the contrary. Today, nine years into the post-Odegard era, the institution thrives as it did in the glory days.

Certainly the tone is different, some of the old stalwarts are gone, the aviation business has changed, and the airplanes are newer and better equipped than ever. But the students still come eagerly all the way to North Dakota, from every state in the country and every continent on the planet, for a chance to become part of the burgeoning world of aviation and space. John Odegard started as a crop duster, at a time when such pilots had to be part daredevil to survive and to thrive. That job is now known as crop spraying, or aerial application, a suggestion that the once nascent world of airplanes has been tamed by the years. At the same time, aviation has gone from being a course of study unheard of and unwelcome at a university to a not atypical part of a higher education curriculum.

At Odegard's school, however, there is still a sense of the crop duster in the air. Call it a feeling, a buzz, a whiff of electricity in the atmosphere. One sees it in the students, marked by energy and serious countenance, rushing from classroom to airport to their part-time jobs. They seem more focused than a typical college student, possessed of a confidence, maturity and sense of responsibility one might associate, perhaps, with a medical student.

"Aero students are my best employees in the store," says Grand Forks City Councilman Hal Gershman, owner of the Happy Harry's chain of liquor stores. "It's been like that forever. Reason? They are very responsible. I sell a product that, if misused, can be very harmful. They don't misuse it because if they get in trouble, they don't become a pilot. They are good thinkers, and they are creative. And they are very responsible."

Again: This stunning byproduct evolved from an irresponsible showoff who was supposed to end a ditch digger without a hole. Which is why the death of the founder serves as the better jumping off point to this inspiring tale. For there are two central questions to be answered in telling the story of the unorthodox founding and growing of such a unique institution.

Number one: Exactly how does one go about creating such a massive and successful school-cum-business against stiff opposition *from absolutely nothing*?

It's often said that Odegard started out with just two planes. In fact, before Tom Clifford persuaded a wealthy rancher named Ernie Fox to donate two Cessna 150s to the cause in 1968, Odegard had zero planes. He had only a far-fetched plan spun out of a further fetched master's thesis suggesting extensive savings in time and money by flying university officials to meetings around the state and region.

There are key lessons to be learned from studying the Clifford-Odegard method— lessons that can be applied to almost any organizational endeavor. But do not look for a one-size-fits all model of success to be packaged and sold. One obvious lesson from this story that should be broadcast far and wide is that one size *seldom* fits all, not even in baseball caps.

Given the nature of John Odegard's forceful, creative personality, the second question that must be answered is almost a corollary of the first: If such a unique blend of personality and drive led to the creation of this world-known aerospace school, how was it possible for it to survive the death of such a hands-on, autocratic founder?

The answer to those questions was visible to all, in a manner of speaking, on the bright October morning in 1998 when John Odegard was laid to rest in a mausoleum in Memorial Park cemetery in Grand Forks. As pallbearers carried his casket along the walkway from hearse to final resting place, it was not lost on many in that overflow crowd of mourners that Odegard, in the last few weeks of his life, had designed his mausoleum and personally picked out his casket.

To the end, Odegard remained the focused micro-manager, even visiting the cemetery repeatedly to be sure the builders of the mausoleum kept it in plumb with the line of the small hillside into which it was built. Many would remark later that as much as they liked the feeling of being in control, they would never have been able to, or want to, carry out such tasks. Then again, they would say, that was just typical John Odegard.

One thing many may not have understood that somber day was that Odegard had also hand picked the six men who were carrying his casket to his grave. But not hand picked through a family member; each was approached personally by Odegard in his last days, and asked to perform this one last service.

It's the picture of those six men, when examined closely, that becomes the microcosm of John Odegard's accomplishment. Among the six were: a former airline pilot, a former astronaut, three former students from the earliest days of the school and a college buddy who went on to become one of the country's preeminent cancer specialists. Each man represents a facet of Odegard's dream and stands as living proof of not just the success before 1998, but the continuance of the dream in the years after.

Consider those Pallbearers:

Jerry Murray was a sophomore in high school in 1968 when John and Diane Odegard moved into the house next door on Reeves Drive in Grand Forks. The subtle way in which Murray was reeled in over the next two years and hooked on a career in aviation is a classic tale of Odegard the master salesman and recruiter. Once Murray had enrolled in the aviation program, Odegard kept him under his wing, steering him toward a job at Cessna Aircraft. While at Cessna, Murray became a public relations expert and today works from Minneapolis, where he caters heavily to the aviation industry.

Jim Bunke grew up in a small rural Minnesota town east of Minneapolis. His father was in the business of starting up telephone companies in small municipalities across the state; he flew to his meetings in his own Cherokee 140. He often took Jim with him, exposing him not only to flying but to the world of business. Bunke had his heart set on becoming an airline pilot, but Odegard saw the makings of a hall-of-fame salesman in his easy sense of humor and ability to talk friends into anything. Knowing it was hard to find a job as an airline pilot in those days, Odegard guided Bunke to an aircraft sales position at Beechcraft. Today the man Odegard always called Bunkerdoodle, lives in Minneapolis where he is a U.S. sales director of business aircraft for Bombardier Aerospace.

Bob Muhs, a North Dakota native, talks of the early days of the aviation program, where everybody knew everybody and considered themselves like a family. Literally. Muhs used to baby-sit John and Diane's children, and one summer he helped paint Odegard's summer lake cabin. Muhs was also a regular at the many informal student discussions held at the Odegard home, arguing the issues of the day with the dean, known to all as simply John. After graduating in 1977, Muhs was hired in the front office of Northwest Airlines at Odegard's urging. Within a year, hard times at Northwest had furloughed him. Odegard hired him back as a staffer until Northwest got back on its feet. Muhs—always Bobby to Odegard—has been a manager with Northwest for almost 30 years.

Bob Buley was born in Canada and raised in Milwaukee. After serving as an enlisted man in the Navy, he tried college but dropped out of the University of Wisconsin to become an airline pilot. He eventually flew for North Central Airlines, which was absorbed by Republic Airlines in the late seventies. He'd never heard of Odegard or the North Dakota aviation program until the day he gave a tour of Republic to some students from Grand Forks. Odegard later invited him to the school where the two became instant friends. When Northwest bought Republic, Captain Buley became a valuable contact for Odegard and offered him much advice and support on expanding the program. While most students and staff called Odegard by his first name, and usually referred to him as JDO, Buley was also the only person Odegard ever allowed to call him Odie.

Jim Buchli grew up in Fargo and won an appointment to the United States Naval Academy in Annapolis. He served a combat tour in Vietnam as a marine platoon leader and only afterward did he decide to become a marine pilot. From there, he was accepted into the astronaut corps at the National Aeronautics and Space Administration and flew three space shuttle missions, including the last flight of the Challenger before it exploded on takeoff in January, 1986. He crossed paths with John Odegard during his astronaut days and became a trusted advisor and frequently-invited lecturer to the university. Buchli even took the UND Aerospace flag into space on one of his shuttle flights. He became the first person to receive an honorary degree from Odegard's school.

Lloyd "Corky" Everson met John Odegard in 1964 when both were undergrads at UND. Odegard had taken over the nearly-defunct Flying Club and was trying to drum up interest among students. Everson signed on, and Odegard taught him to fly. Later, he went with Odegard to a Grand Forks bank and helped persuade them to advance the club a non-collateral loan to pay its past-due bills. Everson went on to earn an M.D. at Harvard and later became a cancer specialist. He started the Roger Maris Cancer Clinic at MeritCare in Fargo, later settling in Houston. When Odegard was diagnosed with cancer and was treated at the M.D. Anderson clinic in Houston, Everson and his wife Jacque took him into their home on dozens of occasions, joined frequently by fellow Houstonian Jim Buchli and his wife Jean.

The high caliber careers of those six pallbearers, and their continued interest in— and support of—Odegard over the years, reflect the main thrust of his dream: to train and fill the aviation world with sophisticated professionals and to develop relationships that could enhance that dream. It was never simply about teaching someone to fly. As was the case with Bunke, Odegard introduced many of his students to a host of non-flying careers, leading to their rise as aviation industry leaders. Moreover, those six pallbearers typify the broad-reaching network of grads and well-placed friends and industry executives that Odegard created—a la his mentor Tom Clifford—and could rely on to keep his program growing.

Attracting people to his cause was, perhaps, Odegard's greatest talent and legacy. Though in many respects he was a confirmed micromanager, he conversely would hire a proven expert to handle the daily details of running the program so that he could manage the big picture.

"The whole aviation program could have been just the John Odegard Show, and when he's gone it was gone," says Buchli. "But that didn't happen. It was founded well on business fundamentals that met a need, then developed that need along with the solution to fill the need—which solution is still valid today. He had people in leadership positions perfectly capable of continuing and growing. The whole structure was very sound, put together very well. And under new leadership, it's very capable of going forward. On the same vector as John? Who knows. It's certainly on a vector toward success."

The story that follows, then, is a story not only of John Odegard but of his pallbearers and many of his flight instructors, teachers, students and essential role-players who now populate the inner corridors of the aerospace world. So many of the people he drew to him were people who knew how to lead from the middle. They gave their energy and faith to a man whose grandiose vision of the future was also accepted on faith by a battle-hardened marine with a blessed weakness for the good-bad boy.

PART ONE
TO LIVE IS TO FLY

To live is to fly

Both low and high

So shake the dust off of your wings

And the sleep out of your eyes.

Townes Van Zandt

Prior page photo: In the early days of the flying club at UND, Odegard (center) got
a manufacturer to donate a glider which he quickly mastered.

A mythical bird that never dies,
the phoenix flies far ahead to the front,
always scanning the landscape
and distant space.

Lam Kam Chuen

Chapter 1
Phoenix Rising

What is it, after all, that makes a boy want to fly? What if he grew up near an airport? Or lived next door to a crop-dusting pilot with a larger-than-life name such as Buzz Gunn—a wooly character said to have flown secret missions for the CIA? Might not that explain why a boy ends up a pilot? Perhaps, but it doesn't really go to the heart of the question. For it's not merely a matter of a boy being fascinated by airplanes or by his neighbor or thinking he could get the hang of flying and not kill himself in the process. The question doesn't ask why a boy wants to fly an airplane or even to be a pilot. It asks why a boy *wants to fly at all*.

Put another way, why is it that *this* boy wants to fly and *that* boy does not? Of course, no one really asks such questions, and even if they did there's no guarantee that one could ever plumb the inner depths to find precisely why. Who knows the ultimate source of that fiery inner voice that whispers again and again, "You must. You simply must." That kind of fire isn't stoked by the mere sight of an airplane overhead or the embellished yarns of a crop duster, not even to a boy pre-disposed to hear the distant thrum of a propeller and to dream himself into a cockpit. The dream is a good start, of course, but it takes something more to get the sensitive seeds of desire to take hold and bloom into a passion, a need.

There is a particular night in this story of one boy's flight where an answer begins to suggest itself. This was a night sometime in 1948 or 1949, when a man named Truman Odegard sat at the bar of the Elks Club in Minot, North Dakota. Minot in those days was a small city of 30,000 in the north central part of the state. If North Dakota sounds weathered and remote, Minot makes a good poster child for the state. It is home to a windblown Air Force base, a small college and a stalwart citizenry born with earlaps on their souls. Their proud boast: "Minot isn't the end of the world, but you can see it from here."

On the night in question, Truman Odegard was listening to a friend on the next bar stool hawking a new kind of insurance policy. It covered medical costs for a terrible disease currently sweeping the nation. The disease, claiming as its victims mostly young children, was polio. It's not hard to see Truman softening to the pitch; he had three young children at home, and his wife, Clara, was active in the March of Dimes. Years later, the March of Dimes would fund the development of a polio vaccine. An affable sort, Truman had his usual beer and a bump and let the salesman blather on.

Truman Odegard was a man who liked his barley pop. He had a cartoonish round belly to prove it and a pedigree of sorts. Years back, he'd attended the University of North Dakota in Grand Forks, hard by the Red River of the North and the Minnesota border. There he'd earned a reputation as a party lover. In fact, he'd been a sidekick of one Whitey Larson, the legendary owner of "Whitey's," the storied Minnesota hangout for UND students just a few feet across the river and the state line in East Grand Forks. Whitey's and other bars in East Grand Forks took full advantage of Minnesota's later closing hours than those in North Dakota. It's been said that Truman, a bit of a risk taker, got involved with some of the storied bootleg activity in the area. In the late 1920's and 30's, beer and spirits were smuggled across the not-too-distant Canadian border, often ending up in hangouts like Whiteys. But Truman eventually partied himself out of the university, only a year short of earning a degree in electrical engineering.

He was a big, bald, good hearted Norwegian, a jolly charmer liked by all. Yet those who really knew him also talked about a rougher side. As a boy selling newspapers, for instance, he'd had to fight it out many times with his competitors to get the best corner for hawking the daily sheet. And during World War II, when Truman was a bombardier in the Fifth Air Force in missions over the Pacific, he was promoted to corporal five times. Which means he was demoted to private four times. Which means he had a problem with his temper. He was one of the first air crews to work with the newfangled Norden bombsite, an accurate instrument that helped to win the war. But it was so delicate that it required of its bombardier a good bit of time, patience and precision while working in a darkened room, to prep the controls *just so* before a mission. More than once someone, usually an officer, would barge into the dark room unaware of what was taking place. The intruder would flick a light switch, and ruin the adjustment. Truman's typically acid response guaranteed he would never become a sergeant.

After the war, Truman worked as a repairman at Northern States Power in Minot. People across the small town knew him well for his house calls to their washing machines and stoves purchased from his employer. They already knew Clara as a hard-working pioneer, the first woman hired at Minot's Gavin Yards by the Great Northern Railroad. By all accounts, the Odegards were local favorites.

Truman was the engaging story teller, a fairly decent cook and the life of the party. Clara, a little older than her husband, wasn't against partying, but she was the strong and stable half of the duo, the one who held things together. She was known in Minot as a caring, forgiving person, often described as the least judgmental soul anyone had ever met. Some saw her also as a well-intentioned enabler for Truman.

The Odegards lived in a modest house at the bottom of a hill near the Roosevelt Slough, a tributary of the ever-winding Mouse River. There they raised three children. Jim, the oldest, born in 1936, was practical, like his mother, and serious. John, the middle child, born five years later, was light hearted, stubborn and full of charm, much like his father. Joanne, the youngest, born five years after John, was a little of both, and became a frequent harassment victim of her charming, prankster middle brother.

The Odegard house was not far from the Minot airport where the local fixed base operator, a man named Pietsch, ran a crop-dusting service and offered flying lessons. Truman Odegard wanted nothing to do with flying; he'd had enough of it in the war and was determined to keep his feet on the ground. Not so with his son John, who very early fell under the spell of the family who lived next door, a family known as The Flying Gunns. The entire family flew, from Fred Gunn, the father, to his sons Buster, Kinky and Buzz. Buster flew crop spraying sorties out of the Minot airport for Pietsch Flying Service. Kinky had been killed in the war, but Buzz Gunn was very alive and a colorful local character. A reputed soldier of fortune who flew as a mercenary in France and Israel, he was said to have flown for the CIA. Whatever, Buzz was seldom without a shaggy dog tale of his exploits—stories that seized the imagination of young John Odegard.

As a boy possessed of limitless energy, John devoted much of it to his fascination for building things; he never tired of connecting odd parts to each other. He assembled plastic airplanes like all boys do, but he also fashioned together his own go-cart and a soap-box racer. Once, clearly under the influence of Buzz Gunn, he built a wooden airplane scooter and persuaded his playmates to push him up and down the street.

That night at The Elks Club, Truman Odegard gave in and bought the polio insurance policy; one can only speculate how it went over with Clara when he got home. But in less than 60 days, their young son John had fallen ill. Clara recognized the symptoms of polio immediately, and she and Truman rushed him to the hospital. That day, another Minot boy John's age was admitted to the same hospital by his parents, and the two young patients became friends. Both were diagnosed with Bulbar polio, a virus which attacks the brain and shuts down muscles needed for breathing and swallowing. Both boys were confined, from the neck down, in a so-called iron lung, a medieval- looking cylindrical contraption that mechanically assisted their breathing. It was of little consequence to Clara that

John Odegard, standing on the sofa in this view from Christmas, 1942, grew up in a modest home in Minot. N.D., not far from the airport where he would one day learn to fly. His maternal grandmother, (white print dress) the family matriarch, lived with the family. Odegard's parents, Truman and Clara Odegard, are the couple on the far right behind the sofa. Older brother Jim, seated, in striped shirt, became John's protector.

John Odegard built his life's reputation very early on going fast. As a young man it didn't matter what he was driving, a boat, a plane or a sports car like this one. As a mature man he never lost the desire to move at high speed, whether in the air or in his mind.

all of John's medical costs were covered, thanks to Truman's drinking buddy. She was extremely afraid she would lose her son, and she had good cause for such fear. Not long after he was admitted, that other Minot boy with polio died. In later years, John would remember lying paralyzed in his iron lung and seeing his family standing at the door to his room looking terribly sad.

There is no pharmaceutical cure for polio once it takes hold, although patients can—and do—spontaneously recover partial or full use of their faculties. Inexplicably, after three months in the iron lung John's muscles slowly began to heal, and he could breathe on his own. But the ordeal, and seeing his new-found friend die of the disease, had changed him. Before the polio, he'd seldom been seen without a smile on his face. When he came home from the hospital, recalls his brother Jim, he seemed very quiet and down.

By all accounts, John Odegard was never an introspective man. Few who knew him recall moments when he shared deep psychological insights with them; he seemed to be one of those people who acted out every thought or impulse, holding back very little. His attitude toward life was defined almost solely by action. Many of his friends and family believe the suffocating, near-death experience of polio had burned into his psyche an everlasting need to be unbound.

"It didn't take him long," says brother Jim, "to realize how fortunate he was to have survived." Soon, John was back to being not just the old John, but the old John at warp speed, a turbo-charged free spirit. "He moved with a group of kids that was willing to take chances and do a lot of youthful crazy things," says Jim. "I think he did more of the leading than following in those cases. He was always in the middle where the action was. He was like a Fonzy," the colorful character from television's "Happy Days."

Throughout the rest of his life John Odegard seldom spoke of his polio trauma, not surprising, perhaps, for a man who could never admit to any weakness. But even his close friend from his later Minot high school days, Del Rae Meier, a woman who remained close to him until the end, never knew about the polio. Stunned upon hearing of it years after his death, she thought for a moment. "It makes sense," she says. "The only way to hold John down was an iron lung. Maybe that's why he had such energy. We were always attracted to that energy. It was reckless. John lived fast. If things could have been done yesterday, even better. When he was young, things just weren't going fast enough."

Again, the question: why does a boy want to *fly*. Go-carts and soap-box racers and motorcycles and sports cars go fast. In his youth, John sampled all of them. But he didn't become the next Mario Andretti. Why? One person with whom Odegard did share his polio experience was Tom Clifford, the man who became his father-like mentor.

"I think he liked the freedom of flying," says Clifford. "And, you know, he told me he had a very unusual pet in those days. A bat. I just think it was that

John Odegard graduated
from Minot High School in
1959. A popular socializer,
he'd dated every cheer-
leader but one—Diane
Rosedale—whom he
never even met until five
years after high school.
When he did, it was love at
first sight for both of them.

Truman Odegard,
John's father, was
well known and well
liked in Minot, and
a frequenter of local
saloons. Often John
and his brother Jim
had to find him
and persuade him
to come home.
John often remarked
that he wanted
more from life
than his father.

feeling of freedom…"

His sister Joanne recalls the bat, or at least a flying squirrel. She remembers the day at the family's summer camp on nearby Lake Metigoshe that her brother fished the injured animal out of the water. He took it home, fed it and kept it in an oatmeal box until it could go off on its own. He later rescued a pair of baby golden eagles whose mother had been killed. He brought them home as well, raising them until they were ready to fly, at which point he turned them over to a park ranger.

And so it was to the Minot airport John would go during his free time as a boy, using his charm and the Buzz Gunn connection to ingratiate himself with the Pietsches. He volunteered for menial jobs such as sweeping out a hangar or washing an airplane in exchange for a free flight or 20 minutes worth of flying instructions at the end of a week. He was a 13-year-old ninth grader, just a couple of years from his bout with paralysis, when he took his first lesson. He earned his private license while still in high school. Coupled with an impish proclivity to raise hell, being a flier made him a natural leader of the pack.

"He was big, and he always looked so important," says Del Rae Meier. "He had a bold persona always. You could look at a group of guys and know he was leading them. He was always at the center of leading you someplace, and you knew it was going to be a wild ride."

He and his loose-knit gang pulled dozens of juvenile stunts or pranks. They would go out at night, for instance, and pound on the outside wall of a small neighborhood grocery, knocking cans and boxes off the shelves. Today, Odegard's daughter Stephanie recalls how she and her brother John, Jr. always believed from those stories they'd heard that their father had been a teenaged hoodlum. Odegard would usually josh charmingly at such talk and say, "Yes, I was full of the dickens." Like when Clara and Truman went off to the cabin at Lake Metigoshe for a weekend, and, "it would echo through the halls of Minot High School," recalled one classmate: "Party at Odegard's."

If his dickens got him into trouble at home, the consequences never seemed to fit the crime.

"Mom did what she always did," sighs Joanne. "She'd say to him, 'Now don't do that again.' Mom and John had a special relationship because she came so close to losing him. And John was very good at getting what he wanted. He'd catch me doing things I wasn't supposed to do, and then he'd blackmail me. When I was 12, he caught me smoking. He said he'd tell Mom unless I waited on him hand and foot. I did that for a long time. When Mom found out she said, 'That's really dumb. Don't do that anymore.' But it went on until we grew up—until he left home."

The day that John was suspended from school for zapping a student with an electrified chair, Clara and Truman seemed unfazed, even as John's teacher reported the incident to them in tears.

"If that had been my child who did that," says Diane Odegard today, "life would never have been the same. But his parents were all right with it. So he was suspended from school for a few days. They would have said, 'Well, really, he's very smart. He fixed that chair up, he did all the wiring. That's not necessary to be suspended from school.' It was no big thing. They were kind of lenient."

"His mother," adds Del Rae Meier, "was such a gracious, humble person. I always say humbleness is knowing who you are and who God is and not getting them mixed up. I really felt John could probably be called arrogant and not humble."

When the trouble he engineered grew more serious, John often went to his older brother for advice, rather than his parents. Jim would go on to law school and become the respected state's attorney in Grand Forks for 20 years. Though five years separated them, they were close, and Jim did a good bit of protecting of his younger brother.

Like his father, Jim was handy with tools and a competent carpenter. He built a floor for an 8 x 10 wall tent a few feet from the family cabin on Lake Metigoshe. He and John lived in it most of the summer, perhaps 30 steps from the lake. "We'd jump in the motorboat and go off together a lot," Jim recalls. "We did a lot of exploring, things boys do around a lake." One afternoon, they were fishing on the north end of the lake when a black squall materialized in the distance. From it shot a small twister. It raced across the bay toward their boat which, at 7.5 horsepower, wasn't going to outrun it. "What are we going to do?" asked John. Jim pushed his brother to the floor of the boat and draped himself over him. The twister blew across the boat, spinning it like a top, but they came out of it fine.

It was at the lake that John showed his first inclinations toward entrepreneurship. He'd noted a lack of garbage pickup for any of the camps along the lake. In the late fifties, at age 17, he and a friend started a trash pickup service, hauling refuse every week to the local dump for a fee. It was young Odegard's first lesson in the reality of a pilot's life. If you wanted to fly, you needed money to pay for flight time and fuel.

When he graduated from Minot High School in 1959, John wasn't sure what to do with his life. His parents had never pressured any of their children to go to college, or, in fact, to follow any particular career path.

"I was always the kind who was planning and thinking of what I was going to do next, rather than sit and wait for something to happen," says Jim Odegard. "I'd made up my mind that I was going to go to college. I went to Minot State. I knew there would be no financial assistance from home. Father spent his money, and whatever money there was came from mother. There was a bond there between mother and John, maybe because she'd almost lost him. She never questioned anything I did but certainly never questioned what John did. When he got out of high school he went off to college in Colorado. The first thing you know, he's driving a beautiful red Corvette. Mother financed it through the Great Northern Credit Union."

The Colorado experiment didn't last long. Odegard spent about a year studying at the University of Denver. He ran out of money, and his grades weren't good, so he dropped out and came back home. He tried his luck at Minot State but soon lost interest and quit. In the meantime, he'd landed a job with Pietsch's Flying Service as a crop sprayer. Later, in 1963, he took a job in data processing at the Boeing plant adjacent to the Minot Air Force base. There, missile silos were being constructed as part of the now- defunct Minuteman Air Defense program.

Odegard was 22 years old at this point and still not sure what he wanted to do in life. He was quite content with the dangerous work of spraying crops and, in the opinion of Del Rae Meier, he might have continued to fly and work just to pay for his flying without any real purpose in his life. But that summer, Odegard's wild and aimless days disappeared almost overnight. That was the summer that a friend and co-worker at Boeing, young Byron Dorgan, happened to be running an errand to the main office in downtown Minot. There he met a brand new summer fill-in named Diane Rosedale, home between her junior and senior years at Milwaukee-Downer College. It was her first day on the job, and she needed a ride out to the Air Force base. Dorgan, a future United States Senator and key supporter of the Odegard aerospace school, obliged. When they got to the base, he introduced Diane to everyone, including John Odegard.

Ironically, even though Odegard had gone to high school with Diane Rosedale, he'd never met her. Odd, because according to Del Rae Meier, John dated just about everyone in his high school class.

"He liked women," she says. "I didn't know all of them, there wouldn't be enough time to know all of them. I just knew he dated a lot. Almost everyone I knew dated him. *I* dated him. Sometimes he'd have more than one date in a day."

Diane Rosedale wasn't a Minot native. She had joined Odegard's high school class in the junior year. She was born in New York City. Her father was a career Air Force man and the family was frequently transferred. Before coming to the Minot Air Force base, he'd been stationed with his family in Chateauroux just south of Paris. Even today when people recall Diane from those Minot high school days, they talk of her as being from France and how exotic it made her to everyone else.

Exotic, however, worked against her. "She was so pretty," says Del Rae. "Some people thought that was threatening. She was a reserved, quiet person. Smart, beautiful, but she wasn't a hugely popular person." She and Del Rae hit it off, though, and together became members of the cheerleading squad. But even though John Odegard had dated cheerleaders, including Del Rae, he still had never made the acquaintance of Diane Rosedale.

"She just didn't fit the kind of person I knew that he was used to dating," says Del Rae. "Which is why no one was more shocked than I when he started dating Diane that summer. That he'd go out with her just didn't seem like a normal connection. But when he and Diane got together that summer at Boeing it was almost like an explosion."

Almost immediately after their introduction, Odegard invited Diane to go flying

Soon after John Odegard earned his pilot's license in Minot, he found part time work spraying crops in a craft like this one. Years later he still considered crop sprayers the best pilots.

After John and Diane married they moved from Minot to Grand Forks and lived in the hutments—old Quonset huts provided for married students by the University of North Dakota. As John worked toward his undergraduate degree in business, he was intrigued by the nearly defunct university Flying Club.

with him. Luckily, his pilot's license had just been reactivated. Officially, he'd lost it after he buzzed picnic tables in Brainerd, Minnesota, where an FAA inspector named Les Severance had seen the numbers on his wing and written them down.

Diane had never been in a small plane before, and John had never dated a woman who spoke French. By the time the flight was over, a torrid summer romance had blossomed.

"I saw his enthusiasm when we first met," Diane recalls. "He seemed to know more than anyone in the office. All the higher-ups and supervisors were his friends. He knew how to fly. He had a red Renault. No one had a car like that in Minot. He was cute. He just was full of it. He just swept me off my feet."

Odegard took Diane to his parents' Lake Metigoshe cabin and sped her around the lake in his motorboat. "He drove it very fast, and everyone knew he drove very fast," she recalls. "That was a new experience for me. I never knew about a cabin or going to a lake. My parents had nothing like that. Lake life was fun: partying and skiing and drinking and eating. His parents were fun. Truman would make things on the outside stove. It was exciting, and John and I fell madly in love."

By summer's end they'd become engaged, planning a wedding the following June after Diane's graduation. "We had very few practical ideas about what we were going to do for money or furniture," says Diane. "We were just going to get married and John was going to finish school."

When she returned to Milwaukee for her senior year, Diane decided to take education courses with the idea of becoming a teacher to help support John's graduate studies. The only other career plan she'd ever considered was becoming an airline stewardess—as flight attendants were called in those days. Secretly, though, she dreamed of the stage, having been involved in theater in college and high school. She'd once performed a one-woman show of scenes from Broadway plays.

Pushed by Diane, John set his sights on the University of North Dakota in Grand Forks. It was in the spring of 1964 when they came to town to see if John could get into the university. There they met Tom Clifford, officially ending Odegard's restless and aimless Minot days of thunder. Diane was the very studious type to whom education was supremely important. She expected John to do well and, to the surprise of just about everybody, he did just that.

"His life just changed with Diane," says Del Rae Meier. "She brought a whole new purpose to his life. If he could have, he would have just gone on flying in Minot. Diane made him more serious. Serious wasn't even in his vocabulary before then. She grounded him."

Literally. When they'd met, John belonged to a parachuting club and was still nursing a back injury from a bad landing. Diane quickly forbade any further jumping. Without argument, John agreed. He insisted, however, on staying with another dangerous activity: the in and out, down and up, acrobatic flying required of a crop sprayer.

On a clear, windless day, a hard flying crop sprayer could make $250 from dawn to dusk at 25 cents an acre. Odegard remained proud of his background as a crop sprayer, regarding it as the ultimate macho pilot credential. There were few regulations then in the seat-of-the-pants crop spraying world, and fewer regulators around to judge. Often, survival was the main proof of ability. If you could fly well enough to spray crops, Odegard's reasoning went, you could fly just about anything; by extension, you also could handle just about anything. During his career, Odegard had a chance to fly with, and to recruit, many experienced pilots for his program. If a pilot had crop spraying experience, it was almost a given he'd become an Odegard favorite.

In Grand Forks that fall of 1964, the newlywed Odegards moved into university student housing at what was known quaintly was The Hutments. They were little more than surplus Army Quonset huts made over into apartments for married students. While the Odegards found them charming, a Minot friend who'd helped them move down to Grand Forks took one look and turned down the invitation to spend the night, preferring to drive the three hours back to Minot.

Still thinking of teaching, or even a community theater involvement, Diane had started graduate school and taken a part-time job at a department store. But she realized, soon after enrolling, that she and John couldn't afford for both of them to be in college at the same time.

"John was in school, and I really needed to work. His things were very important. Theater wouldn't fit into our life. There was no question I was giving up something. It was just the natural thing you did then. He had to do what he had to do and I needed to work," says Diane.

She quit grad school and got a job as the youth director at the Grand Forks Air Force Base while John enrolled in UND's business school. She was pleased to see that he attacked his studies with his familiar energy; in fact, there seemed to be no holding him back. One thing Diane noticed: unlike Clara and Truman, John was now very channeled and focused. In fact, he was so focused on what he really wanted out of life that it didn't take him long to hear about the disjointed, leaderless remains of an on-campus organization called The Flying Club.

Ostensibly, the club was a student-run organization that leased time on small propeller-driven airplanes and paid for it by collecting flight fees from student members. It had no affiliation with the university and very little standing among other clubs on campus. The club was heavily in debt for missed lease payments and fuel costs. With fewer than five members, it had pretty much ceased to exist. It was at this teetering instant that the Great Playwright in the Sky brought John Odegard onstage. To his unbound mind there was no disarray, only a golden opportunity to turn nothing into something. Odegard saw, in the wreckage of the club, a dormant power base; a skeletal structure of respect and authority. It lacked only someone with energy and vision to raise it from the ashes, much like the phoenix of mythology, and make it fly again. And, of course, the ability to get a million dollars wouldn't hurt either.

A rock pile ceases to be a rock pile
the moment a single man contemplates it,
bearing within him the image of a cathedral.

Antoine de Ste. Exupery
Flight to Arras

Chapter 2
An Excess of Enthusiasm

Long before John Odegard, North Dakota had already produced a handful of colorful aviators. The most notable was the World War I pilot and explorer Carl Ben Eielson. A native of Hatton, North Dakota, he'd served in the Army Air Corps during The Great War and in 1921 graduated from the University of North Dakota. He then became one of those legendary risk-taking air mail pilots written about so gracefully and poignantly in books such as *Flight to Arras* and *Wind, Sand and the Stars* by the French aviator Antoine de St. Exupery. These were still the dangerous pioneering days of aviation, and Eielson made a name for himself by becoming the first pilot to fly air mail to Alaska. While in Alaska he became the first pilot to over fly the North Pole and later made history with dramatic flights across Antarctica. As with John Odegard, the tragedy of Eielson's death at a young age—crash landing in Siberia while attempting to rescue the crew of a sinking steamer—adds a mixture of sadness and glory to his exploits.

Don Smith, a man whose unique aviation industry experience was to make him invaluable to John Odegard in coming years, adds to North Dakota's flying lore with his yarn about an aviator named Elroy Bollinger. A Navy veteran of the first world war, Bollinger taught industrial arts at the University of North Dakota. In 1925, two years before Lindbergh flew non-stop to Paris, Bollinger decided to have his class build and fly a stick glider. They bought a kit from Jack Northrop—the former chief engineer at Lockheed who would go on to found his own aerospace company (which, today, is Northrop Grumman.) Northrop was so interested in Bollinger's glider project that he took a train to Grand Forks to inspect the assembly before the glider was flown.

The scene of the flight was a field on the campus at the end of an old trolley line—just about where the university's law school stands today. Bollinger climbed into the cockpit, the glider's nose attached to the rear bumper of an old Willys night car with a 200-foot length of rope. With Northrop and a small crowd of the curious looking on, a student volunteer put the Willys in motion. Billowing smoke, it trudged across the field, hauling the glider into the air. As the rope tether fell back, the glider soared into the sky, giving Bollinger a unique North Dakota aviation first.

As an historical footnote, Don Smith asserts he knows this to be fact because he heard it directly from Bollinger himself. "I met him for the first time in New York City," he recalls. "His daughter Ann went to college with me. He was then Dr. Bollinger—director of industrial arts for the state of New York. Although I really didn't get to know him until I eloped with his daughter."

Given such claims to fame, it isn't surprising that a Flying Club arose at the university. In fact, during the Second World War, the military called on several colleges to help train a civilian air corps to feed the ever-increasing need for pilots. It paid the colleges for all training costs, even building small airfields for some of them. After the war, some of those colleges—Ohio State, Purdue, Southern Illinois—re-developed the military program into an academic course. Some, like the University of Auburn in Alabama, already had an ongoing aviation curriculum. Most such programs were ad hoc add-ons to either an engineering school or a business school.

In 1947 a National Intercollegiate Flying Association, or NIFA, developed. A sort of trade association for college programs, its mainstay was an annual flying competition where schools sent their flying teams to compete in a series of flight tests that ultimately crowned a national champion. The lure of such competition inspired other colleges to create fledgling programs. In truth, it usually wasn't the college per se that had the vision to start the programs, but an enthusiastic faculty member with a keen interest in flying.

Someone like John Odegard, for example. In 1964, he didn't even have a college degree, but it didn't stop him from throwing all of his energy into rejuvenating the dormant Flying Club in Grand Forks. He packed an important credential that enhanced his ability to attract members and to save money: He was a certified F.A.A. flight instructor and could train student pilots without having to farm out the work to the more expensive fixed base operator at the Grand Forks Airport.

The first thing Odegard did when he easily assumed command of the Flying Club was to prioritize its needs. Which was worse: A flying club so in debt it couldn't rent or lease airplanes? Or a flying club without members to fly the non-existent airplanes? A subtle but important first decision: Odegard decided if he had people they could help him raise money for aircraft. So he turned on his full powers of persuasion and, as he had done to everyone in Minot, quickly attracted a following. He'd do this by staking out a spot in the middle of the busy Student Union on campus. He'd put up a sign that said, "Join The Flying Club" and grab people as they passed by.

"He'd be out there slapping people on the back, talking them into joining the club," recalls one of the earliest Flying Club joiners, Don Johnston, "A lot of students just wanted to belong to something, to have friends. A lot of them didn't fly much—it was a big expense. But everyone loved to belong to it, because John made it a fun organization. There was something going on every week—airplane washes, raffles, parties. We had some great parties."

Today Johnston is a senior 747 Captain at Northwest Airlines, and one of Odegard's very first graduates. He remembers in the early winter of 1964 being dragged by a friend to a Flying Club meeting and meeting Odegard for the first time. The club had about 15 or so members at the time. Johnston, who'd always had an interest in airplanes, went back for the next meeting and was impressed that Odegard remembered his name.

"He wasn't fakey, like some people," says Johnston. "He had a way of making people feel like 'I'm living for you, I am so glad to see you.' He made you believe him. When he dreamed, everyone else dreamed with him."

As soon as students signed up, Odegard gave them jobs to do, errands to run, questionnaires and surveys to write up and distribute—all aimed at building support for the club. Members eagerly did anything he asked. "The only reason it got started was the strength of John's personality, his drive, his passion," says Lloyd "Corky" Everson, one of those lured into the club by the sign and the sales pitch. "He knew how to sell an idea and then execute on it."

From the beginning, Odegard ran the club like a business. He created an executive board that held monthly luncheons in the student center. Members who were assigned a task were invited to the meeting to give a report and, most importantly, to get a free lunch. "It was great experience," says Johnston. "It was part of our training in how to take on a leadership position."

The money for those lunches came from the $5 monthly dues paid by each member, plus a separate billing for flight time on the club's leased planes. Odegard charged members $6 an hour to rent a club plane, undercutting the local fixed base operator who charged $12 to $14 an hour.

Odegard found the money to lease the planes in the first place with a series of fund-raising gimmicks such as weigh-ins, where anyone could get a plane ride by paying so much per pound. He also went off campus and solicited club memberships from merchants in Grand Forks. The university rule was that a club member had to have an affiliation with the university, but Odegard got around that in clever ways. Dave Vaaler, for instance, a local insurance executive, was allowed to pay a hefty membership to join the club simply because he attended UND hockey games. Odegard found a way to allow just about anyone to join, and he created dizzying levels of membership status from active, to active associate, to associate, to inactive associate and whatever else he could dream up.

Within two years, by 1966, the club's membership had grown to 200 members, and its treasury suddenly had plenty of cash. Odegard had solved the club's immediate debt crisis by simply going downtown to a local bank and asking for a loan. He took with him two club members, David James and Corky Everson, and laid out in enthusiastic detail how the loan would be repaid—through fees and memberships.

"I don't know what the bank must have been thinking," recalls Everson today. "We didn't have any collateral." But Odegard seemed to have it all together, and it probably didn't hurt that he dropped the name of Tom Clifford at every opportunity. Bernie McDermott, vice-president of the First National Bank, gave them their loan and got the club back in the flying business. "We signed the notes," says Everson, "and that was the start of John's efforts to build this into an amazing concern."

According to a tongue-in-cheek story by former North Dakota governor Allen Olson—who became one of Odegard's avid supporters in the early 1980's—the eventual success of the entire aerospace program hinged on a decision Olson made as a law student at UND shortly after the Flying Club got its loan.

Olson's father, a North Dakota rancher, owned a small prop plane, and he frequently took his son with him on cattle buying trips. Worried that he might become ill while flying, he'd taught Al to fly so that he could land the plane in an emergency. While Olson was in law school in Grand Forks, a friend proposed that they share the $300 cost of buying an old Taylorcraft prop plane, known then as a T-Craft. Olson's father offered to lend him $150, but before Al committed, he wanted a test flight. On takeoff he noticed oil leaking back onto the windshield from the engine. So he passed on the deal. His friend, however, found another partner to buy a share—a Flying Club enthusiast named John Odegard. In later years, when he was governor, Olson liked to kid Odegard that if he hadn't passed on buying that T-Craft, the Flying Club would have fizzled, and the great aerospace center would never have happened.

From the moment Odegard went back to college he maintained close contact with Tom Clifford. He had an edge over other students who might also have wanted the ear of the dean: Odegard could fly, and Tom Clifford hated the frequent four-hour drives he had to make from Grand Forks to the state capitol in Bismarck, where he often testified before the legislature. By now Clifford had been appointed vice-president for finance by UND President George Starcher and was the official voice of the university in defending the annual budget. It wasn't long before Odegard was flying both Clifford and Starcher back and forth to Bismarck in planes leased by the Flying Club.

"John had an advantage," says Don Johnston. "Starcher and Clifford loved to take the plane out to Bismarck. That gave John an hour and fifteen minutes with a captive audience. A lot of things were accomplished on those flights."

Clifford was impressed with the initiative Odegard had shown in getting the Flying Club in such good shape in a short time—all while completing his

By the fall of 1968, helped by the support of Tom Clifford, then a UND vice-president, Odegard won permission to start a department of aviation in the business school. Later that year John and Diane bought their home on Reeves Drive in a neighborhood of stately older homes and spreading elm trees.

After Odegard resuscitated the UND Flying Club, one of his first recruits was Lloyd "Corky" Everson, whom John taught to fly. The Odegards spent a lot of time with Corky (far right) and his wife Jacque (second from left), even vacationing together here in Mexico. Everson became a renowned cancer expert who would help Odegard in his later struggle against the disease.

undergraduate degree. After Odegard earned his B.A. in 1966, Clifford admitted him to the graduate school where he pursued a degree in accounting and worked as a data processing assistant. It was during those hops to Bismarck with his captive audience that Odegard began pitching ideas to Clifford about formalizing the unofficial flights to the state capitol. His idea was to start an air service that would fly university officials and professors anywhere in the state or region they needed to go. The way he envisioned it, the university would reimburse the Flying Club for all costs of such trips.

He pushed the idea hard, and Clifford later would laugh that Odegard always broached the topic when he had his passengers a little bit vulnerable at several thousand feet in the air. Still, Clifford encouraged Odegard, who made the air service idea the subject of his master's thesis. His argument showed it would be cheaper in the long run to fly officials to state meetings than to have them drive and be reimbursed for mileage, meals and hotel bills. It was cheaper, but only by a small margin. From November of 1967 to the middle of the summer of 1968, for example, the Flying Club made 85 trips ferrying faculty and staff to state meetings. For those trips, the university paid the club about $8,700. The university comptroller, Gerald Skogley, said the university saved about $2,300 or about $27 per trip. Odegard would continually make the argument that if one factored in the value of time wasted in long automobile trips, the service was more than worth it.

Another item Odegard pressed while high above the prairie was the idea of starting an academic course in aviation. While Clifford was encouraging, other university officials were less so. Bill Konker, the academic vice president at the time told Odegard bluntly that he had an "excess of enthusiasm." He was the first, though certainly not the last, to ask of Odegard when he persisted with his idea, "Just what part of no don't you understand?"

As Odegard later explained to Dick Youngblood of the *Minneapolis Tribune*, "They didn't just say no, they said hell no—never in fact."

But Odegard talked up the aviation course with his Flying Club members, and even advanced the notion of an aviation department. It would be a tough sell, even if Clifford and people like Bill Konker agreed. A new academic course would have to be approved by the fiscally conservative state Board of Higher Education, no small obstacle. In the meantime, as a kind of jump-start to the whole idea, Clifford gave Odegard permission in 1967 to teach a ground school class. This was a basic introduction to flying that covered aerodynamics, equipment, tower procedures, pre-flight checks, weather and instruments—in short, most of what a pilot needed to know about flying short of actual flying. It was a low-overhead program, but it drew a full load of students.

Along with his ground school, his data processing job and the work for his master's degree, Odegard was also using Flying Club members to research the steps required for the club to buy its own airplane. "We were just students doing this,"

In high school a counselor despaired that John Odegard would ever amount to anything, not even a ditch digger. But encouraged by Diane, he earned a business degree in 1966—a cause for Truman and Clara to travel down from Minot. Truman, once a student at UND, had dropped out. John would earn a masters degree a year later.

Odegard reorganized the university's unofficial Flying Club, took out a bank loan to pay past debts and recruited scores of new members. The club leased planes and Odegard, a certified flight instructor, gave lessons, with fees paying the club's bills. Here, Odegard and an unidentified club member pose with one of the club's early Cessna 150s.

says Don Johnston, "but we went about it like a corporation. We'd sit down and say, 'How do you buy an airplane?' We'd say a plane costs this much new, so we'll amortize this. We'd figure our hourly expenses and see if we could justify how much to charge per hour. It was a great educational experience."

Odegard even farmed out the flying of lesser university officials across the state to other members of the club—members he'd personally taught to fly.

When Odegard finished his master's degree in 1967, Clifford hired him as an accounting instructor. The dean had personally approved his master's thesis and, along with President George Starcher, was leaning in favor of starting up a formalized university air service. But Odegard wanted more. He continued to lobby Clifford and Starcher for an official aviation program beyond the ground school. Odegard had done his homework. He noted that most of the aviation world was being run by World War II veterans who were starting to retire. He predicted an industry need in coming years for graduates who were well-grounded in business. They would become the next generation to run the airlines and general aviation companies.

In anticipation of that, Odegard had started his own survey of curricula of other colleges with aviation programs. At a 1967 flight competition at Parks College in St. Louis—sponsored by NIFA, the organization of college flying programs—Odegard began introducing himself around to the directors of various college aviation programs. He wanted detailed information about their curricula, but from most received a polite suggestion that he consult their course catalog. Later in the day, he shook hands with a man named Gary Kiteley who ran the aviation program out of the engineering department at Auburn University in Alabama. This would be one of the more fortuitous hand shakes of Odegard's career.

The Auburn professor asked him where he was from. When Odegard said North Dakota, Kiteley couldn't resist a horse laugh.

"I said 'You gotta be kidding,'" Kiteley recalls. "I was in Minneapolis in aviation for four years, and I left Minnesota to go someplace warmer. North Dakota was the last place I'd consider for a flying program."

Odegard smiled and shrugged. "Well, that's where I am."

He told Kiteley of his unusual position as an accounting professor-unofficial university pilot-ground school leader. He said he wanted to start an official aviation program. He asked Kiteley if he would help him put together a curriculum. Entranced by Odegard's brashness and energy, Kiteley agreed to send Odegard the detailed course outlines for the entire Auburn aviation curriculum. Thus began a solid, 30-year friendship that would produce long-term benefits for both Kiteley's and Odegard's programs.

In the meantime, the short-term gain was Odegard's ability to recite to Clifford and Starcher chapter and verse on just what an aviation program would look like at

Now that he had a ground school, Odegard went looking for a curriculum on which to base classes for a full department of aviation. At a collegiate air show, he met Gary Kiteley, who ran the aviation program out of the engineering department at Auburn University in Alabama. Kiteley gave him the Alabama curriculum and later the two became national leaders of college flight-education organizations.

In 1967 Tom Clifford gave Odegard permission to start a ground school as a preliminary to opening a department of aviation. Would-be pilots learned the basics of flying and weather and mechanics. One of Odegard's students was Bruce Smith, a UND football all-American who would one day succeed Odegard as dean.

North Dakota and what it would take to get it going. In fact, the program that eventually started was modeled almost completely after Kiteley's; over the years, Odegard missed few chances to pay tribute to Auburn University for its generous support.

Even with a strong footing for a program, the idea of starting an aviation curriculum faced enormous obstacles. It wasn't merely the state Board of Higher Education that had to be convinced. The traditional, hidebound faculty at the University of North Dakota had to be won over; that seemed about as easy as convincing a mime to talk.

"Our biggest hurdle," says Clifford, "was that the faculty didn't believe aviation was a university subject. There was a certain amount of snobbery involved. They saw flight training as more or less a vocational thing. Training a pilot was like training someone to drive a truck. What they failed to capture was that this was an entirely new industry. There was a whole lot more to it than training pilots."

Odegard made no secret of his disdain for the so-called academic, and it hurt his cause more than once over the years. In the tie-dye days of 1967, the standard garb of a college professor had devolved from suit and tie to the de-rigueur counter-culture uniform of jeans and a beard. Odegard branded them all with the term "sneaker creeps." They, in turn, talked disdainfully of "teaching monkeys how to fly."

But, in fact, Odegard had staked out an academic philosophy for aviation education. His position: the excitement of the cockpit was just a hook to induce students into a program that gave them a complete liberal arts education and also a solid business footing in a growing field that currently lacked gravitas. Only four years earlier, Odegard was content with a life as an aimless college drop-out and part-time crop sprayer. Now he was flying much higher, advocating aviation study as a way to professionalize the world of flying. He argued that those who entered an industry as polished professionals would end up running that industry.

For all his disdain of academics, Odegard seems to have quickly grasped a vital reality of university life: Send forth into the world graduates who will become successful, and they will remember their alma mater generously; the alma mater could then send forth even more students who will keep the cycle alive. A cynic bent on empire building might espouse such a philosophy, but so also might an idealist desirous of improving the commonweal. John Odegard was often accused of empire building—usually by those who sniffed that no one was building *them* an empire. And as a businessman to the core, Odegard often displayed more than a modicum of cynicism. But never enough to overpower a boyish idealism—a redeeming trait of his that often goes uncommented on. For idealism was very much part of the equation that set his grand plan in motion.

By his own admission in later years, Odegard had been neither a solid nor a serious student. Whenever he talked of his own schooling, he did so in a tone tinged with apology and chagrin. He would also tell people in later years that he'd never had a teacher in school who made an impression on him—except to tell him what a failure he was destined to become. Yet during his brief career as a married undergrad and then grad student in Grand Forks, Odegard appears to have experienced an awakening. Some of it, no doubt, is due to the focused cheerleading of his wife, Diane, and the fatherly interest of Tom Clifford. But these were also the years when Odegard himself became a de facto mentor to the eager members of his Flying Club, often only a few years younger. Odegard had personally recruited all of them, nurtured them, taught them to fly and to be responsible. In one way or another he was always inside their lives.

"He'd just show up at the right time when you needed it," says Jim Bunke. "I used to use the expression that I really believed in him. By being around him you just felt better about yourself. He saw something in you that you hadn't seen yet. He saw potential in the least likely cases. He did a wonderful job with the guys who might just need a little boost."

While flying had become Odegard's pathway to outward freedom as a young man, the power of teaching a lesson and seeing it take hold in an eager student seems to have fed an even deeper emancipation. Education suddenly took on a status in John Odegard's heart very close to that of flying. For a man who knew only one speed—All Ahead, Full—he became a passionate, supremely confident proselytizer for this new religion. He was hard to resist.

"I can't imagine anyone other than John convincing Tom Clifford to promote that aviation program," says Gerry Skogley, then the university comptroller. "And Tom would not have promoted it if he didn't think he could make it legitimate."

Clifford might simply have pushed an aviation program into his business department, but that wasn't his style. He was a master of what today is known as consensus building. In 1967 it was still a matter of schmoozing, horse trading and three-card Monty. When the university's curriculum committee opposed the addition of a program on aviation, Clifford said fine. He came back with a proposal for something he called "The Economics of Transportation." The committee had no problem with that and approved it; the dean and his upstart pilot shared a chuckle.

"I thought it was valid," says Clifford. "We'd had a survey made that showed the airlines were going to need thousands of pilots. The industry was going to be burgeoning and growing. The country couldn't do without it. If you were going to train someone for a C.P.A., why not train them for the aviation industry as well? It was ambitious for us, but we were interested in growing and being the best in the field."

That wasn't the end of the political hurdles by a long shot. But it was enough to get a program started. In the fall of 1968, Odegard began teaching the first course in what would one day become a college of aerospace studies named after him. Clifford found him a tiny office on the ground floor of the new—and appropriately named—Gamble Hall which housed the college of Business. Odegard packed the closet-sized room with dozens of hanging model airplanes and photos from aviation history. Only four years removed from buzzing picnic tables, the freshly-rehabilitated Odegard settled into a freshly-invented title. He was now director of the brand new UND Aviation Program, specializing—nudge nudge, wink wink—in the economics of transportation.

As for the future,
your task is not to foresee it,
but to enable it.
Antoine de Ste. Exupery
The Wisdom of the Sands

Chapter 3
Changes in Attitude

From his second floor bedroom window at the back of his family's home on Reeves Drive, high school sophomore Jerry Murray had an unobstructed view of the house next door. Every morning around 7 a.m., as he got dressed for school, he would watch curiously as Mr. Odegard, his new neighbor, emerged from his back door and headed for his garage. Mr. Odegard, he knew, was a pilot and some kind of big shot at the university. Just watching him leave the house and walk down the driveway, get into his car and leave, *just by the way he carried himself,* Murray knew to a certainty that whatever he did at the university he must be successful at it.

Sometimes when Murray got home from school late in the afternoon, Mr. Odegard was pulling into his garage. He'd spot Murray and step across to his yard, striking up an animated conversation. Murray would ask, "Did you go anyplace interesting today?" Usually Mr. Odegard, who insisted that Jerry call him John, had just come back from flying the president of the university to Bismarck.

"Holy cow, that's great," was Jerry's standard reply. Even though he didn't know a Cessna from a Boeing, Jerry knew that you couldn't drive to Bismarck and back in a day. Yet here was his neighbor back home already. He liked the way John treated him as an equal and that you only had to ask him one question, and he'd tell you everything about his day.

"He never had an air about him," recalls Murray. And Odegard never tried to push him into aviation. That is, overtly. For instance, he never offered to take him flying, never offered even to take him to the airport. There was one time, however, when he did ask Jerry what he wanted to do in life. Jerry wanted to be a wildlife biologist. John listened and dropped a small stone in the well. "You know," he said, "wild life biologists that are pilots get the best jobs."

Jerry had no interest in becoming a pilot, yet he did feel captivated by his neighbor's charisma and energy. "He had me thinking. And I'm thinking, 'How am I going to hold off this person who is trying to influence me?' He didn't say I shouldn't major in wildlife biology. But if I wanted to get a good job, I ought to be a pilot. And I'm thinking 'How am I ever going to not end up learning to fly living next door to him?'"

The answer was simple. There was no way.

"I didn't realize it at the time," says Murray, "but as I look back on it I can see that when John Odegard moved in next door to me, my course was set."

Perhaps it was the maturing effect of parenthood that polished the edges on John Odegard's recruiting technique. By 1968, when he and Diane moved into the house on Reeves Drive, a traditional Grand Forks neighborhood of stately older homes and spreading elm trees, they had a one-year-old son, John, Jr. A daughter, Stephanie, was on the way. The days of standing alone in the student union under the Flying Club banner and doing an instant hard sell on anyone who walked by were gone. As were the days of the tin hutments, when, strapped for cash, the Odegards amused themselves by inviting couples such as Corky and Jacque Everson over to play charades. As it turned out, Odegard was a master at charades.

One of the ironies of John Odegard's life—the unchained boy who saved bats and eagles and longed to soar above the mundane—was the staid, anchored-to-the-ground career choice he nearly made instead of flying. It wasn't just that he came to Grand Forks in 1964 to get a college degree; he came to become an accountant. While earning his degree, he drew interest from several accounting firms around the country. In those hutment days, Diane and John did a lot of talking about which big city they would move to once he got his accounting degree and C.P.A.

This, even though during graduate school John had told Diane—following an audit he'd conducted at the North Dakota State Mill and Elevator—that he was never going to do that kind of work again. He was good at numbers, he understood and was savvy about the way of business, but he hadn't the patience for a black-and-white life based solely on columns of figures. Many years later, during one of the regular professional retreats Odegard would hold for his staff, he required everyone to take the Myers-Briggs personality test. Everyone got a chance to compare their score with their colleagues' to learn more about them and ultimately work better together. Terri Clark, a C.P.A. who has become the long-time director of fiscal affairs at the aerospace school and the Aerospace Foundation, was among those taking the test. She recalls being shocked to find that Odegard's score put his personality type deep into the dreamy, idea-prone stylistic categories and as far away from someone suited for the highly-organized, meticulous rigors of accounting as the test allowed.

"I said to him, 'How did you ever get to be an accountant?'" Clark recalls. "He told me then that he didn't like accounting. He said, 'Oh I hated it. But I had to have something just in case.'"

Still, says Diane, "I don't remember ever talking about an aviation department then, or starting a school. He probably had it in his mind, but he never talked about it." Nor did he discuss with Diane ahead of time his decision to reject the offers of the big accounting firms. "I thought we were going to go to other places," she remembers. "But he said to me one day, 'It probably isn't relevant to keep saying we might go here or there...because we're doing this now...'"

Odegard did most of his career planning with Tom Clifford. The UND vice-president had clearly taken Odegard under his wing and persuaded him to turn down the accounting jobs to work temporarily as an instructor at UND.

"He was okay as an accountant," says Clifford "but it was never his passion. Still, I was an accountant—sometimes a little more daring than most—but I didn't wear green eye shades and armbands. We had a kind of a kindred soul in that sense."

Clifford promised Odegard that if he stayed at UND he eventually could have a small aviation department. "I was not particularly interested in aviation," says Clifford, "but I could see the future of it and see where it ought to be developed." Thus, almost as soon as Odegard was ensconced in his tiny Gamble Hall Office, he and Clifford embarked on an aggressive campaign to twist official arms in favor of creating an aviation department.

Odegard flew Clifford on multiple trips across the state where they dropped in unannounced on each member of the state's Board of Higher Education to argue their case. The dog-and-pony show even extended to Odegard's loyal following in the Flying Club. Don Johnston recalls that he and others often flew supportive faculty members to Bismarck to help persuade legislators of the importance of an aviation department. Board members and legislators were impressed with Odegard's enthusiasm and endorsed his vision. In early 1969, when Odegard officially qualified for his C.P.A., his ad hoc aviation program became an official department in the College of Business. Clifford raised his protégé to the status of chair and promoted him to assistant professor.

Odegard continued to wear several hats. He served not only as chair but taught several classes and took his turn as a pilot in the university air service—even flying emergency medical cases from the surrounding area to tertiary care hospitals in Minneapolis-St. Paul, or Rochester, Minnesota.

Having an aviation department and keeping it, however, were two different things. The devil's deal Odegard had made—that he would never ask the state or the university for funding—defined his approach from the start. The department was to be run not so much as an academic program as it was a business enterprise. For his needs were clear: increase the enrollment so that more students and their tuition and flight fees would be on hand to feed the department kitty. At the same time, increase the size of the fleet to handle more students with more tuition and fees.

From the early days, Odegard knew that students who could afford the expensive fuel and training costs of an aviation degree—even today it is not unusual for an aviation major to graduate with a student debt approaching $45,000—weren't all going to be found in North Dakota. From day one, Odegard saw that his department would have to become national in scope to attract enough students to keep it afloat as a going concern.

That's one of the reasons he had attended that meeting of NIFA at Parks College in St. Louis, where he met Gary Kiteley from Auburn and got a blueprint for a curriculum. He'd also introduced himself at that meeting to dozens of general aviation industry executives, such as Cessna's Russ Watson. It made sense for Cessna to ally with organizations such as NIFA and the University Aviation Association. Collegiate flying programs, though still fairly new in the late sixties, purchased a lot of airplanes. Watson was a marketer but also headed Cessna's educational outreach program. He knew just about everybody in the college aviation world and proved an invaluable asset to Odegard.

"John had a very high regard for education," says Watson. "He loved his students. All you had to do was just be around him. It was a joy to walk through his office with him. He'd introduce you to everyone, from students to a new flight instructor he'd hired to Tom Clifford. He had tremendous pride with everything to do with the program. And almost everyone in it had been to his house and knew his son and daughter and wife. Even if you were a visitor, he made you feel that you belonged."

Odegard often cemented such relationships with gifts. Anyone paying a visit to his department went home with several bags of Red River Potatoes, courtesy of Tom Ryan, one of Odegard's new-found local friends. Not only were those potatoes a hit with every visitor, it was said that Ryan, the self-made millionaire potato farmer, would hand pick the biggest and best for Odegard's special guests.

"Just before I'd leave to go back to Wichita," recalls Watson, "I'd load the plane up with potatoes. I'd usually bring a sack back for Russ Meyer, our president. I'd give them to my neighbors, my minister. Everyone out there had a high regard for Red River Potatoes."

Within a year of meeting up with Kiteley and Watson, Odegard was a well-known figure on the national collegiate flying scene. Soon, he and Kiteley began talking about taking over both NIFA and the UAA. The two organizations were then run by a revered veteran flier named Harold Wood, of Parks College. According to Kiteley, Wood was approaching retirement. Kiteley and Odegard thought they could revitalize both organizations to spark further interest in aviation among the nation's high school students and thereby eventually benefit their programs.

In the late sixties, Kiteley got himself nominated to the NIFA board. He began working with fellow member Russ Watson, of Cessna, to expand the organization into multiple regions. At the time, NIFA sponsored just the one annual intercollegiate flying competition, along with a mid-winter business meeting. The expansion would create regional competitions and regional interest while at the same time making the national convention more prestigious.

Odegard, in the meantime, gravitated toward the UAA. In 1969 Russ Watson was a member of UAA's nominating committee; he approached Odegard and asked him to become president. "Everybody was his friend," says Watson. "Truly. This position let him develop some very nice associations."

"I think the University Aviation Association was a means for John to understand what the competition was doing," says Bob Muhs. "He watched his flanks. This was a means to make sure his department had a national standing and at same time a means to ensure they weren't missing any kind of competitive edge. He never ignored that. He absolutely hated to get beat by anything or anyone."

Not long after he assumed the top UAA post, Odegard and Kiteley had discussions about standardizing the curriculum of collegiate aviation programs. Odegard chaired a committee that developed common nomenclature and academic credits for certain courses. Kiteley chaired the accreditation committee that would develop educational standards.

Their joint efforts culminated in a first-of-its-kind workshop held in Wichita, Kansas, often called the home of general aviation because so many industry giants were founded and are still headquartered there. The workshop, chaired by Kiteley, was held in 1976 and hosted by Beechcraft and Cessna. The result was the publication of the first-ever college aviation training standards manual, including several contributions by Odegard. The manual is still in print today.

On the home front, in the meantime, Diane Odegard was enjoying the results of getting John to finish his education. Not only had their financial picture brightened, but their social life had picked up steam. Grand Forks, like many small communities, had a reputation as a closed place; it was very hard to break into. Soon, however, the Odegards had joined a church, John had become a member of Rotary and Diane was asked to join the board of the Grand Forks symphony. John became friendly with several doctors, one of them Rod Clark who had delivered the Odegard's first child, John, Jr., in the fall of 1967 and daughter Stephanie two years later. Clark would later buy two Cessna aircraft that he leased back to the school.

After Stephanie was born, Diane went back to teaching and was even thinking again about getting her master's degree. She enjoyed the mandatory social obligations of being the wife of the chair—the formal white glove receptions for the dean, worrying about saying the right thing as she was introduced to faculty members. "It was very exciting because nobody's husband was a pilot," she says.

But there was a downside to success.

"John would never get home on time," Diane remembers. "He was always late in my opinion."

Typically in those first several years, he was either hiring someone or arranging for a world expert in some aspect of aviation to visit or to run a seminar. Or he'd be attending a committee meeting or training himself to fly the latest airplane in the growing fleet. Surrounding all of it was the never-ending stressful reminder, says Diane, that "You have to be continually making money. Because there's no money coming from the university and no money coming from the state. So you've got to be your own business. And it got bigger and bigger and bigger. And John went more and more places."

Diane soon realized that it wasn't only in a motorboat or a red Renault or an airplane that John Odegard moved fast. "He was always rushing," she says. "I wanted him to be home, and he wanted to go back to work. He had a young family and two young babies. There were some times when there was stress, but basically John didn't change at all, except he got busier and busier."

It seemed as if there was always some meeting or trip to take him away from the family. "At first I thought that was not good, until I realized that these trips had to happen. At first I was, 'Why can't you be home mowing the grass?' We'd be having a neighborhood party or a picnic, and he wasn't interested at all in that kind of stuff. I realized he couldn't be home mowing the grass if he was going to be working on this other stuff. So I mowed the grass. Everybody else's husband was home mowing the grass. He was off in someplace like Russia doing very exciting things. I just had to go with it. Then, as things got more interesting, his trips got more interesting. And I would vicariously go on the trips with him. So I remember changing my attitude."

Jerry Murray also remembers a day when he changed his attitude. It was in August of 1971, the day before he was to enter the University of North Dakota as a freshman to begin studying wildlife biology. He was in his backyard one afternoon when he heard his name called. It was John Odegard, standing at the fence that separated their houses.

"You know where you're going in the morning?" he called.

Actually, Murray didn't. The idea of going down to the huge campus and figuring out which building to go into was daunting.

"Not really," he said.

Odegard gave him a typically short reply, one that went straight to the heart of the matter. Murray would soon come to recognize it as a familiar Odegard refrain, one that always proved correct.

"You'll figure it out," he said. "Don't worry about it."

From the day John Odegard (left) moved into the house next door on Reeves Drive, says Jerry Murray (right), his future was sealed. Odegard subtly recruited the eager young Murray into the aviation program. He became a pilot for Beechcraft and now consults out of Minneapolis in aviation-oriented public relations. Here, Jim Moore, a colleague of Jerry's (center) mugs with Murray and Odegard.

For the first 15 years of its existence, UND's flying program was headquartered in this cramped office in Gamble Hall. Here, as chair of the department of aviation in the school of business, Odegard spent time "cooking up schemes" that not only built up the fledgling aviation program but invented an entire department of atmospheric sciences.

Odegard never asked him if he was learning to fly or if he had chosen aviation. His style when recruiting a student in those early days—and there was no doubt he was recruiting Jerry Murray—was to work one student at a time. So Odegard spent some time telling Murray about a goose hunting trip he'd just come back from with Tom Ryan. Murray was an avid hunter and hung on every word.

He finished his story and said, "Well, what are you majoring in?" Murray told him wildlife biology.

"Well, you know," said Odegard, "wild life biologists that are pilots get the best jobs."

As before, that was all. Nothing further transpired for four months, months during which Murray continued to replay Odegard's words. In December, the day before his first semester's final exams Murray went down into Gamble Hall and found Odegard's office.

"I didn't know I needed an appointment," he remembers. "I didn't even know how to make an appointment. I walked in and started to ask if I could see John. He sees me from his open office. He comes bounding out and says 'Jerry, come on back here.'

"So, now, this is a classic scene. You get to his office. You sit there while he makes multiple phone calls. And takes multiple phone calls. And you're just fascinated being there, watching and listening to him operate. I only had one question. And I'll bet he kept me waiting an hour while he's talking nonstop on the phone. He's talking to politicians and deans and vice presidents and people in industry. He knows you're there. But you found out right away that if he could be on the phone with someone and was advancing his program that was what he would prioritize. He spent his life on the phone. He was always on the phone.

"So he finally hangs up, and he's off the phone. He goes, 'How was your day?' I say it was good. He goes, 'Um, did you come to see me?' 'Yeah, I did,' I say. 'What kind of jobs do aviation grads get?' He goes, 'Good jobs.'

"And the phone rang, and he picked it up. And I thought I've got to get home for dinner. And the next semester I enrolled in aviation 101 and started flying. It didn't dawn on me until after I graduated that he was the consummate recruiter. He knew how self-confident I was. That I really believed wildlife biology was the right major. Yet he was influencing me. Recruiting me. Later, I watched him recruit other people and saw that if he decided he wanted you, he'd never stop trying to recruit you. What I learned in school was that everything he did was for the kids, all those phone calls were about building the program for the benefit of the kids."

If you want to make enemies,
try to change something.
Woodrow Wilson

Chapter 4
What Color is Your Airplane?

While Tom Clifford helped deflect many of the political objections to an aviation department, there were several practical hurdles Odegard needed to overcome to make his department viable. The biggest, of course, was lack of funding. The university owned no airplanes and wasn't disposed to buy any, no matter how much President George Starcher and his vice-president for Finance thought of Odegard's idea. Still, an aviation program simply had to have some sort of fleet.

That's when, in 1968, a character named Ernie Fox entered the picture, almost *deus ex-machina*. Fox owned land in Lakota, in the north central part of the state, not far from Clifford's family homestead near Langdon. Clifford knew Fox well. As a boy he had often hunted duck and goose at a small pond on his neighbor's property. Fox lived most of the time in Montana and had purchased the oil rights to a good bit of land both there and in North Dakota. Clifford had once visited Fox in Montana and had been amazed to pick seashells out of the bluffs on his property. The land had once been covered by ocean and now, eons later, the undersea vegetation had corrupted itself into oil, bubbling right there through Ernie Fox's soil.

Fox was well known for his distrust of lawyers and accountants, of which Clifford was both. But their long friendship overcame such small change; Clifford did Fox's income taxes every year as a favor, never charging a fee. One day in the summer of 1968, he got Fox on the phone in Montana and called in a marker. "I said, 'Ernie we need airplanes. We know where we can get two for the price of one. Will you finance one?' And he said, 'Yeah.' And he did."

Those two Cessna 150s were soon joined at the Grand Forks airport by an old wooden Mooney Executive that the Flying Club had financed through the First

National Bank, leasing it back to the university by the hour. For some time, it became the mainstay for the university air service. Odegard also managed to get several local lawyers and doctors like Rod Clark to buy airplanes and lease them back to the university.

But even with this handful of planes, a more serious problem developed. North Dakota's fixed base operators (FBO)—those people who ran the general aviation side of things at local airports, and who rented out planes and gave flying lessons— saw Odegard's flying of university officials around the state as unfair competition. They saw the idea of the university teaching students to fly as grossly unfair competition, and they began to howl.

One of the most vocal FBOs was a man who had once employed Odegard in the mid-sixties as a crop sprayer and flight instructor. Jim Montgomery had been the FBO at the Grand Forks airport since 1946 and was said to have trained dozens of local pilots up and down the Red River Valley.

"Jim was a short, stocky Irishman who'd been a Golden Gloves boxer," recalls Don Johnston. "He was a great wit, he told a good story, but he was very belligerent."

Montgomery and Odegard were very close at one point, but their friendship evaporated shortly after the aviation department got started. Until then, the university had always contracted with Montgomery to provide flight training for students enrolled in the Air Force's Reserve Officer Training Corps program on campus. In 1968, when Odegard's aviation department opened, the nation was in the midst of a massive troop buildup for the Vietnam War. The Air Force found itself running more students than ever through its ROTC the program. In the first flexing of his opportunistic muscle, Odegard convinced the Air Force that his program was better suited to handle student training. They ended up giving him the contract.

It was a move that cost Montgomery a good bit of business. He immediately felt betrayed that Odegard would compete against him after he'd showed him so many tricks of the pilot's trade.

"He also resented that university faculty were hopping rides with the Flying Club," says Don Johnston, "instead of paying him about four times the cost along with the price of hiring a pilot." Visitors say they often heard Montgomery mutter, "I'd be better off if I had never instructed one pilot."

Letters to the editor of the *Grand Forks Herald* from Montgomery began appearing at that time, stirring talk against Odegard's new venture. In those letters, Montgomery expressed anger that the city of Grand Forks—which owned the airport at the time—required that he, as an FBO, maintain certain standards such as having a heated shop and a mechanic on duty at all times. He even had to post a bond of $50,000—requirements from which Odegard's program was exempt because it was part of an educational institution.

In the late 60s, a Texas aviation company donated several used DC-3's like this one, to college aviation programs, including UND. Odegard used his to fly faculty and university administrators to Bismarck and to carry the hockey team to its away games. When several airborne engine failures were labeled sabotage, the FBI investigated. Odegard had many enemies among jealous faculty and fixed-base operators, but no one was ever arrested.

From the beginning, Odegard stressed safety in his flying program. He insisted on thorough maintenance of his department's aircraft. Often he would show up at the airport and inspect the equipment himself. He wouldn't hesitate to order new parts and even new planes, even though his department often had no money.

Even so, Odegard's first university operation at the airport was pitiful. Its administrative offices were housed in a double-wide house trailer. The airplanes were kept in a war surplus hangar whose door wouldn't shut, letting in all the elements from those wind-blown winters.

Montgomery wasn't Odegard's only critic. "Guys in the flight school business or charter aircraft business all hated John," says Jim Bunke. "The guy who owned the flight school in Devil's Lake hated him as much as the guy in Valley City. Their interpretation was that he was using state dollars to compete with small town operators for flight students. There was a real disdain for John. The crop sprayers even disliked him."

Such criticism, which bothered Odegard immensely, stirred up the antagonism of some state politicians who didn't like the university to begin with. It might have faded quietly with time if not for a series of disturbing incidents that began taking place at the airport not long after the new department opened.

In the fall of 1968, Odegard heard about a millionaire from the southeastern part of the country who wanted to donate an old Douglas DC-3 twin-engine prop plane to an appropriate college. Supposedly a deal had been struck whereby a college in the south agreed to take the plane. But after lengthy bureaucratic bumbling, so the story goes, the unnamed college demanded of the donor that it be delivered bearing its own colors. Outraged, the millionaire retracted his offer.

Odegard heard the story and tracked down the donor.

"I understand that you were willing to give a DC-3 to a college," he said over the phone.

"Yeah," said the voice on the other end of the line. "And the sons of bitches wanted me to paint it in their colors."

Odegard said he'd be interested in the plane and the donor replied sarcastically, "Oh? And what are North Dakota's school colors?"

Not missing a beat, Odegard replied, "What color is your airplane?"

While that may or may not be an apocryphal story—it is often told as an example of Odegard's hubris—the millionaire in question did not deliver the plane to UND. But shortly thereafter, another one did: Harry Bradley, president of Houston Aviation products of Houston, Texas donated several used DC-3s to various places including the University of North Dakota.

The 26-passenger plane, delivered to UND in October of 1968, was immediately pressed into service, flying the university's hockey team and basketball team to their out- of-state games. It was also used to fly larger groups of university officials to Bismarck and back. To pilot the plane, Odegard hired a veteran Minnesota crop sprayer named Bill Knox. He'd flown C-46s and DC-4s for cargo airlines and had spent ten years as the private pilot for an electronics firm in Northern Minnesota. One credential that landed him the job was his experience flying "tail draggers"—planes with a third landing gear in the tail as opposed to the more modern tricycle-style rig featuring a wheel under the nose. The DC-3, of course, is a tail dragger.

Knox found the plane easy to handle; because of his experience flying in bad weather across Minnesota, frequent flights through snowstorms didn't bother him. Things went well with the plane for about a year. Then, on Christmas night, 1969, as Knox was flying the university basketball team back from a game in Iowa, he noticed a fluctuation in the oil pressure and oil temperature gauges on the right engine of the DC-3. He immediately shut it down and, flying on one engine, made it safely to Minneapolis. After a careful ground examination, it was determined that a deteriorated bearing had caused the problem.

The plane was repaired and put back in service. Knox had never had an engine fail on him in his entire career and figured this was just the odds catching up to him. Three months later, however, while taking off from Bismarck with President George Starcher, Tom Clifford and John Odegard all on board, the same engine blew a cylinder.

As it happened, Starcher, only months from retirement, was at a window seat over the right wing. He looked out and saw fire shooting from the engine but seemed completely unbothered. Afterward, when Knox had managed to get the plane back onto the ground, Odegard asked the president if he had been worried. Starcher seemed surprised and said no. He explained he'd been on a night flight once out of Minneapolis and the plane had gone through some clouds. He looked out the window and saw fire coming back from one of the large exhaust pipes on the engine. He'd nervously called the flight attendant and told her the engine was on fire. She laughed and told him that was just something known as "torching in the clouds," a phenomenon that occurs when the heat of the exhaust gives the impression in cloud cover of flaming along its edges. Therefore, he told Odegard, when he saw the flames coming from the engine this time, he was determined not to say anything.

As for Knox he was grateful for the years of training that made wrestling the bulky plane on one engine seem somewhat second nature. Nevertheless, he considered himself lucky to have put the craft safely on the ground. "You do what you have to do and afterward think 'Boy, I'm glad that's over.'"

A lingering worry remained: he had now experienced two engine failures in a matter of months after years of never before having such a problem. "After a while you're scanning this panel of gauges all the time," he says. "Just automatically, in a quick glance, to see if everything is where it's supposed to be. If one moves out of where it's supposed to be, your eye picks it up. Otherwise, if you didn't see it, the engine might seize and the crank would break and the propellers might go flying off."

Odegard, too, was concerned and had the plane thoroughly examined, but nothing was found. Once the costly engine repairs were made, the DC-3 was put back in service.

Three years later, in late March of 1973, Knox was getting ready to fly the university hockey team back to Grand Forks from Duluth. It was late in the afternoon, in the midst of what was described in the newspapers the next day as a "blinding snowstorm." Knox and his passengers were on board and waiting for the de-icing to be completed when Rube Bjorkman, the hockey coach, poked his head into the cockpit. He shouted that the plane couldn't take off until 4:15 p.m. It was then about 4 p.m.; Knox said fine and idled away for several minutes. Bjorkman then came back and gave him the high sign. Knox asked him the reason for the delay. The hockey coach explained that everyone on the team had put a dollar into a kitty and bet on the exact minute the plane would take off. Bjorkman chuckled that he had bet on 4:15 p.m. Knox shook his head, released the brake and rolled down the runway into a seething blizzard.

The DC-3 had just cleared the runway and was perhaps 100 feet off the tarmac when something on the gauges caught Knox's eye—another odd fluctuation. "I was just so lucky I saw it," he says. A second later the right engine blew out. The minimum control speed for the DC-3 was 96 knots. At that moment it was climbing at only a few knots above that. But with all power in the right engine suddenly gone, the manifold pressure down to nothing, the dead engine became a gigantic drag on the liftoff momentum. Its propeller began windmilling dangerously. The DC-3 was only seconds away from sinking like a stone onto the runway.

The standard drill in such a case calls for the pilot to instantly pull back on the prop of the dead engine and to perform what's known as a feathering of the propeller—yanking it out of its crazy pitch and bringing its blades back to a streamlined configuration. Immediately four words raced through Bill Knox's mind: throttle, prop, mixture, feather. In such split-second moments of stress, it's happened that a pilot will react in confusion and pull back on the wrong prop. Knox quickly hauled back on the prop of the right engine and by dint of the noise level staying the same, the plane still behaving the same, he knew with instant relief that he'd pulled back on the correct propeller.

But even as he hit the feather button Knox could almost hear his instructors from long ago haranguing him that you never turn into a dead engine. Easier said than done: When a motor quits like that, the remaining good engine pulls the plane in the direction of the one that stopped. It's one thing to lose a bit of altitude from, say, 3,000 feet while you're bulldogging the plane around to the left. But the DC-3 had already described a 45-degree tilt to the right, and Knox and his co-pilot were making no progress in fighting it back to an even keel, let alone getting it hauled over into a reverse arc. He estimated that he had perhaps 100 feet of ceiling above him to go along with the 100 feet of altitude below. He remembers noting almost absently that the heavy snow being whipped against the plane wasn't sticking.

It was then that a pilot's instinct, bred from years of experience, quieted those voices from the past. "I thought, 'I'm just going to let this plane turn me around [to the right] and get me lined up on the runway," Knox recalls. "When you're only 100 feet up, you don't have much room."

As the plane continued to come around to the right, toward the dead engine, air speed fell to 97 knots. Knox risked dropping the nose to gain some speed and to keep the power of the good engine from rolling the plane completely over. He was very close to being out of control and knew it.

"I'm just lucky there wasn't anything in the way," says Knox today. "Somehow we were able to drag it around and put it back on the ground."

The entire emergency took less than a minute, one of those minutes that lasts a lifetime. As the plane landed, Knox heard enthusiastic applause coming from back in the coach. He quickly realized it wasn't from players grateful for the skillful piloting that had just saved their lives. Oblivious to the near catastrophe, hockey coach Rube Bjorkman had jumped to his feet and hollered exuberantly, "That counts as a takeoff!"

By now everyone was concerned that something was seriously amiss with the DC-3. Three engine failures in a matter of months on a plane that had been thoroughly checked out raised ugly suspicions. Samples were taken from the seized engine. Analysis showed the presence of antifreeze, which contains sugar. Sugar, as any pilot knows, gums up an engine and can cause it to abruptly stop working. All of which raised the possibility of sabotage.

The Grand Forks police and even the FBI were called in to investigate. Jim Montgomery complained in another letter to *The Grand Forks Herald* that the FBI had questioned him. He denied any involvement.

The DC-3 was repaired but over the next few months it experienced three more engine failures, and three more successful landings by Bill Knox. Antifreeze was again found in the damaged engines. Because the aviation program had no secured hangar at the time nor a full-time airport crew, it was difficult to keep the plane under watch to prevent someone from tampering with it. Eventually the costs of replacing the blown engines became too much, and the DC-3 was sold. Authorities never made an arrest, and the mystery was never officially solved.

"We had our suspicions who it was," says Knox. "But we couldn't prove it. Every time I took it out, I'd sit on the end of runway wondering, 'If the engine quits what am I going to do?' That could have killed the whole hockey team or basketball team."

Over the next three decades, Odegard's program would face almost continuous assault from political and academic critics who resented him for one reason or another—most of those reasons traceable to jealousy and to Odegard's obvious contempt for those not attuned to his vision. While some of those critics went to great lengths to make the going hard for Odegard, nothing as serious as sabotaging a plane filled with students ever occurred again. But the DC-3 incidents assured that any future criticisms of the aviation program were overloaded with tension, suspicion and bitterness. An "us against them" attitude quickly developed on campus between Odegard's program and the rest of the university.

Years later, one of Odegard's brightest students, Jean Haley Harper—who would go on from UND to become the first woman captain at United Airlines—offered a poignant postscript to the Jim Montgomery piece of the story.

"When I got there in 1971," she said, "I saw a big rift between the established guys at the airport and what they called the 'U.' The U was not well liked. The scuttlebutt I heard was that Montgomery hated John. I talked to John, and he'd say, 'Jim, what a wonderful guy. He was honest and good to me. I like him.' Something wasn't lining up. I wondered what John had done to hurt him. So I introduced myself to Jim, told him my Dad was a crop duster. He said come over and have coffee. So I went over there several times. One day he looked out and said, 'I remember before there was a tower here. And then it got real busy. Then I had to get radios for all my planes. Now I have to taxi around these 727s.'

"And I thought that was it. Change was hard for him. Then he looked over at the big hulking hangar the aviation program had and he said, 'You know I used to be able to watch the sun come up. I could sit here and feel the sun warm me up in the morning...'

"Then he just looked at the hangar, shook his head and didn't say anything. He was lamenting the change. He didn't hate John. It was making his business harder to maintain. He wanted things to be the way they used to be and they never would be again. It was the bottom line for everyone's irritation with John. He was changing things. I found out Jim Montgomery was a very decent man. John was truly, visibly shaken by the situation. It was very sad."

Chapter 5
The Piano Man

From the beginning, John Odegard fell into easy friendships with the elite of Grand Forks. Tom Ryan, the North Dakota potato king and avid big game hunter took a liking to Odegard and made him a frequent partner on his safaris. Ryan's largess on behalf of the aviation department would go far beyond free sacks of potatoes over the years. Then there was the Grand Forks insurance executive, Dave Vaaler, who insured the Flying Club's early air planes and continued to underwrite the program over decades to come. Typically, Odegard got close enough to Vaaler to convince him to take flying lessons and get his pilot's license—which he did. From there, he got Vaaler to buy a pair of airplanes and lease them back to the university where student fees would pay off the loan.

"He had a great personality," says Vaaler, "that made it easy for him to get to know people and to get them to respond to him and get enthusiastic about what he wanted to do. He got them to kind of set aside the 'I don't think you can do that John.' It was a lot of fun but a financial mistake to get involved in airplanes. You can't use them enough to make it worthwhile unless you've got some kind of business that needs them. The things I got into with John that cost me money were my own fault. He got me excited about doing it, and I bought those bananas. I could never feel anything but friendship and admiration for him."

In 1970, Odegard met a most unusual member of the Grand Forks elite, so unusual that most of the city's *intelligentsia* had no idea he was one of them. In fact, Don Smith was that rare outsider who had recently moved to North Dakota not to seek his fortune, but to use the one he already had to repay what he considered a moral obligation to the state.

Smith, as it would become apparent, was cut from the same bolt of lightning as Tom Clifford and John Odegard. He was in his late forties, a creative, self-made man who, as it happened, had an extensive background in the aviation industry. But he kept a low profile as he continuously worked at re-making himself. At the time he met John Odegard, Smith and his wife owned and operated a modest bungalow-style motel on South Washington Street not far from the university. In the meantime, he served quietly on the Grand Forks Board of Adjustment and let it be known that he would volunteer his time to help any business having financial difficulties.

Before they actually met, Smith and Odegard had heard bits and pieces of each other's story from a furniture store owner in town named Rob Larson. It was Larson who outfitted Don Smith's motel. When he learned of Smith's extensive history in aviation, he told him of his close friend John Odegard and of his ambitious aviation plans.

"He told John about me and told me about John," says Smith. "I thought maybe John was smoking something. And I figured John probably thought I was 300 years old and the biggest bullshitter who ever lived."

Indeed, his story rings like fantasy, starting with Smith's ancestors who, he says, settled in New York in 1629 on a farm given them by the King of England. Today the farm is known as Lower Manhattan. Smith says his father pitched for the Philadelphia Athletics (confirmed in the MacMillan Baseball encyclopedia) and later studied automotive engineering at the prestigious Pratt Institute in Manhattan. That is where Smith himself graduated with an engineering degree after serving in the Navy in World War II—he was wounded on Iwo Jima, the same island where marine captain Tom Clifford earned one of his three Pacific purple hearts.

Before the war, as a teenager, Smith was a professional musician, playing piano and organ. He organized several big bands and jazz groups and played at the storied music rooms of the Edison and Taft hotels. As a boy, his father's best friend was a pilot, a local barnstormer who instilled in the younger Smith a love for aviation. He remembers often whenever his parents went out for an evening, taking the drawers out of a cabinet and climbing onto the middle shelf and pretending it was an airplane.

After the war and his graduation from Pratt, Smith was ready to move to Michigan to enter the Chrysler Institute for a master's degree as an automotive engineer. Before that could happen, he had eloped with North Dakotan Ann Bollinger, Elroy Bollinger's daughter. As they were soon expecting a child, Michigan was put on hold.

At about that time, when he and Ann were living on Long Island, Smith happened to see a Lockheed Constellation—a four engine aircraft with a distinctive triple tail—landing at a nearby field. Introduced in the summer of 1939, the "Connie" had been taken over by the military during the war but taken back by Lockheed to become a staple for airlines everywhere. Curious, Smith went over to take a closer look. He met people at the airstrip and talked himself into a job.

Lockheed, a California company, was setting up a brand new service division on the east coast, in order to sell the Constellation and to train pilots. Young Smith was placed in charge of building temporary construction sheds. He did such a good job that when the actual facility was built, he was named plant engineer. He was 25.

Soon Lockheed decided to build a Constellation base at what was then Idlewild airport. "The company was proposing a couple of very conventional hangars," Smith recalls. "There was no way, using standard hangar design, they were going to meet their requirements. I thought the ideal hangar for us would be a cantilever design, one with no columns at the door opening. It would be very expensive to build. They'd have a hangar a mile long with no columns at the door opening. But structures were my favorite subject in college. I came up with the concept on a napkin in a Greek diner outside of Idlewild airport. I then did a stress analysis of the structure. It surprised me, and I worked this thing out. It could cut $780,000 out of steel costs."

The hangar was built according to the design on his napkin, making it the first cantilevered hangar in the world. Smith, in fact, was invited around the world to talk about his design, and Constellation customers from across the globe came to the opening of the new terminal. The notoriety of the hangar brought him to the attention of Lockheed executives who quickly sent him to Harvard to study management. After that he became a regular part of the company's inner circle.

He eventually left Lockheed in the late 1950s to start his own innovative engineering company. He'd taken a hard look at the new generation of civilian aircraft being produced in the post-war boom. He realized that ground equipment like the standard farm tractors used to haul older planes out of their hangar were no longer adequate for the larger jet planes. His company blossomed in its design and manufacture of a range of modern ground support equipment. Ten years later, when he sold his business, he was named executive vice president of the new parent corporation—American Avitron—running five of its companies.

At about this time, in the late sixties, the Smith's daughter, a student at the University of North Dakota, fell ill. Doctors couldn't determine what was wrong. Eventually a UND psychologist worked with her, and she recovered to resume a normal life. The Smiths were so grateful that they tried to buy the doctor a new car.

"But," Smith recalls, "he said, 'Oh no, this is part of the job.'"

Still, Smith felt a moral obligation to somehow repay North Dakota for helping his daughter. He quit his lucrative job, and he and Ann moved to Grand Forks. They opened their motel mostly for fun and began looking around for ways to help. (An editorial aside: It's very likely that for some people the story about the King of England giving Smith's ancestors lower Manhattan is more believable than the idea that a well-off millionaire would sell all and become a management missionary in North Dakota. In Grand Forks, though, where faith has always been strong and pretensions low, nobody had any doubts.)

When they eventually met in 1970, John Odegard could hardly believe his luck. Smith could hardly believe Odegard.

"I met John and I thought, of all places, North Dakota. Maybe if it were Chicago or L.A. or New York it would still have seemed ridiculous what he wanted to do. He wanted to tie together an in-depth professional flight program with a four-year degree in business. Normally people who like flying don't like business. John had a couple of planes that belonged to doctors and two that Ernie Fox bought, thanks to Tom Clifford. Our challenge was 'How do you build a business in aviation which has a healthy appetite for capital, with no money and no support?'"

Smith himself, supplied the answer. "I didn't need an income," he says. "John got together a team of individuals who had already established themselves as winners in their professions and they could work for little or no money because they had other income. That was the key to it all."

When Don Smith signed on for long hours at no pay, it increased the staff of the Department of aviation by a third. Odegard had already hired Elton Lee Barnum, a retired flight instructor and maintenance specialist from Fargo. His background proved as varied and fascinating as Don Smith's.

Born in Brooklyn, Barnum attended the Julliard School of Music in Manhattan where he studied the violin. A natural singer, he taught himself to play the piano, the organ and the trumpet and, according to his wife, Lee, appeared on the stage of the Metropolitan opera.

After leaving Julliard, he became a Presbyterian minister. He met his future wife, Lee, an acting student, while he was teaching philosophy and religion at the University of Evansville, in Indiana. By then he'd become a pilot. He later joined the Air Force as a Chaplain. Following his military years, Barnum became a rural pastor, earning the title "The Flying Parson of South Dakota." On Sundays, he and his wife would fly to seven different remote prairie churches where Barnum would conduct services. He became discouraged, says his wife, by what he called "Sunday Christians," who ignored their faith through the week. He eventually became a full-time flight instructor, teaching in Sioux Falls, Rapid City and, in the late sixties, in Fargo. That's where John Odegard found him and persuaded him to join his Department of aviation.

"They got along well," says Mrs. Barnum. "They respected each others' talent. John was always asking Elton about safety matters. Elton had a true love of flying. He logged 50,000 miles and never had a pilot-caused accident. John wanted to keep his students from having accidents, so they worked well together."

Both Barnum and his wife had an interest in theater and for years were involved with local drama organizations. Barnum was also the theater manager for the Grand Forks Symphony. Barnum became Odegard's chief flight instructor at the airport and taught ground school classes on campus.

"John was the expert in aviation academics," says Robert Reis, an early Odegard financial advisor and later Director of the Aerospace Foundation. "But John didn't like a lot of detail work. It slowed him down too much. So for detail and excellence in education he looked to Lee Barnum. Lee was a guy who just wouldn't accept anything else in a classroom but perfection. John liked that. He wanted our students to be perfectionists. He wanted them to be the best in the field, so when they left here they would be recognized as the best. And those who wanted the best would send their sons and daughters here to school."

While Barnum may have been a perfectionist, the early airport facilities he taught out of were anything but perfect. He and a small maintenance crew worked in an old barn, actually more of a super-sized Quonset hut that once had been the main hangar at the old city airport five miles east. It was picked up and moved in 1968 when the new city airport opened. While better than nothing, the aviation department didn't even have full use of the barn.

"We had a quarter of that old city hangar," Don Smith recalls. "The city put a new sheet metal roof on it, but they lapped it the wrong way and it leaked. When it rained it poured inside the hangar. On top of that, there were birds in there all time. The doors wouldn't close. You had to push 'em open and closed with a tractor, and they got damaged. They were really bad working conditions."

Though Smith had been "hired" to teach courses and to help Odegard administer the department, it wasn't long before the one-time Lockheed executive was out at the airport washing windows, cutting the grass and trying to fix the leaks in the roof. Barnum, Smith and volunteers helped drape a tarp around their quarter of the hangar to keep it warm in the winter. But the wind still whipped in and blew the tarp aside. Some old heaters were set up inside but they didn't do much for the hands—keeping hands warm was crucial to performing maintenance on airplanes. The standard operating procedure that evolved: put on a pot of coffee—not to drink, but to hold on to and keep the fingers warm.

When the aviation department was organized in 1968, the university ordered that fees collected from student flying be used only to offset the cost of ground school or classes directly connected with flight. That's how flight instructors were paid. But what of the academic courses taught back on campus—airport management, airline operations and air transport utilization? That's what Smith wanted to know. He found out that there was absolutely no money to pay for such instructors, and yet the curriculum had already been approved and students accepted into those courses. Smith began to sense he'd come on board in the nick of time for he ended up teaching all of those courses—a total of 13 credit hours a week.

"I said to John, 'Where's the syllabus?'" Smith recalls. "He said, 'You have to make it up.'"

It wasn't the only surprise in Smith's first days on the job. To convince him to join the department, Odegard had promised Smith a private office. When Smith got to Gamble Hall, Odegard stood outside his own office and pointed to the open office next door.

"That's going to be your office," he whispered.

"But John," said Smith, "there's a guy sitting in there."

Odegard nodded, keeping his voice low. "As soon as I figure out how to get rid of him."

Turns out the occupant of the office was a retired history professor who had been assigned the room as one of his retirement perks. In the meantime, Odegard showed Smith to his "temporary" office, an empty 6' x 6' closet. Smith sighed and went down to the basement in Gamble Hall looking for a desk. The only one he could find that would fit through the door of his closet was one of those rickety typist desks, old, shaky, the wood completely dried out. Back upstairs, Smith asked Odegard what about air in the closet? How was he supposed to breathe? No problem, said Odegard. He'd put in a work order to have louvers installed in the lower half of the door.

It would be four long months before the history professor disappeared.

"Don Smith was an interesting man in his own right," says Earl Strinden. "He was an engineer, an entrepreneur, a wonderful piano player, a collector of old Packard automobiles. A very talented man. John was always flying pretty high, and it was important to have someone grounded like Don to make sure it was all fitting together."

That's also the take of Cedric "Tony" Grainger who helped start and head up a new meteorology program in 1980 when Smith was still an important part of the school. "A lot of the program wouldn't have been developed if it hadn't been for Don standing behind John," says Grainger. "He did an awful lot of the work."

Grainger's colleague, Leon Osborne, agrees. "Don did the heavy lifting. John was fast paced. Not a detail person. He was the big picture guy. Don took care of the details."

One of the most important of Smith's duties was to run interference for Odegard with the rest of the university. That started right in the School of Business.

When Tom Clifford became president of the university in 1970, Clare Rowe took over as dean of the business school. One of the things that made life difficult for Rowe was the lack of academic standing of members of the aviation department. It wasn't just that Odegard didn't have a Ph.D.—he always had an insecurity about not having the ultimate academic credential—but that he and his small department were watering down the ratio of Ph.D.s within the business school. To maintain accreditation, the business school needed to have a certain number of Ph.D.s within each discipline. Aviation, so far, had none. In fact, there was no such thing on any American campus as a Ph.D. in aviation.

Many wonder what would have
become of Odegard's program
without the help and wisdom
of Don Smith. A former
Lockheed executive and
inventor of the first
cantilevered hangar, Smith
and his wife moved to Grand
Forks to be near their daughter.
He and Odegard bonded
immediately. Smith become
Odegard's right hand man.

Odegard got into hot water when he leased a corner of this old Grand Forks Airport
Quonset from the city in 1968 without university permission. His program gradually took
over the whole barn—seen here in the early 70s. Eventually a modern training complex of
hangars, offices and a tower developed over the years, still anchored to this old barn.

That made it doubly difficult for the program to gain acceptance with the rest of the faculty on campus. They also didn't like Odegard's frequent end runs around faculty rules and tradition. Whenever he faced a problem, it seemed, Odegard went straight to Clifford.

Faculty quickly grumbled that the new president was showing Odegard favoritism. Aviation students, who took their liberal arts courses with the rest of the student body, were often referred to by professors as "Johnny's boys" or "part of Odegard airlines." Many grumbled that aviation was a technical pursuit and properly belonged on the campus of the state's vocational college in Wahpeton.

"Universities," says Earl Strinden, "are very much creatures of tradition, of governance run basically by committees made up of folks who want to make damn sure nothing changes. Most are not used to the rough and tumble of the competitive world or dealing with someone like John who never thought of himself as an academic."

Still, Odegard took the faculty's criticism of his department personally. Ultimately, says Don Smith, it served as a great energizer.

"The people who were against him motivated him more than anything, to prove that they were wrong," he says. "There was so much negativity. I always called the faculty 'the forces of evil.' It's one thing to disagree, that's fine. Sometimes it's very healthy. But this was bitter. I would go to an occasion somewhere, and it was as if I had a disease. They resented the fact that John had Tom Clifford's ear."

Odegard often directed Smith to attend various faculty meetings on his behalf. "He always managed, conveniently, to be out of town," says Smith. "I always went for him and would sit there and feel kind of all alone. John didn't have an awful lot of patience with the faculty. He was a misfit. Here we were, a company, essentially, with no funding, operating under the umbrella of the university. At these meetings they'd be discussing who was going to get the wooden desk and who got the metal desks, and here we were trying to figure how to survive."

The critics of the program weren't limited to the faculty or to disgruntled FBOs. Legislators, from time to time, began grousing about the aviation program. That kind of problem, though, was something Tom Clifford was expert at handling. Early on in 1968, a local businessman in Bismarck, a political maverick who had run unsuccessfully for governor, seemed to make the death of the aviation department his pet project. The businessman had led many referral campaigns over the years—referenda that would force the state legislature to stop funding a particular program. He'd even tried to refer the entire state university budget, but the state Supreme Court had ruled that the legislature had to fund those institutions.

The critic took out newspaper ads and paid for TV commercials attacking the university's use of public funds for an aviation program and for airplanes like the DC-3. Like many of the program's critics, the Bismarck critic misunderstood the

As soon as Odegard signed the lease for the city's old airport
barn—without any authorization—he slapped the university's
name on it, lest there be any second thoughts. The sign greatly
irritated the local fixed-base operator at the Grand Forks airport,
who saw Odegard as unwelcome competition.

actual financing of Odegard's programs. The DC-3 was a gift and, beyond that, the university did not fund Odegard's department. Yet to critics Odegard seemed always to be spending money and bringing in high-powered people to unfairly compete with local aviation companies.

Indeed, Clifford had funneled some university funds to Odegard. And there was still some doubt in his mind about whether the whole program would succeed. The Bismarck critic's public campaign simply added to the pressure Clifford felt about keeping the program going.

"There were times when Tom had to be frustrated," says Don Smith. "When Tom was dean of the college of business, he liked the idea of the aviation department and supported John. He was part of the team. But as president, he had to go down the center of the line. Still, something had to be done about the Bismarck situation."

In typical Clifford fashion, he met the problem head on. He made an appointment with the critic in Bismarck—a man whom he'd never met. The only thing he really knew about the man was that he was Irish, like Clifford. Odegard flew Clifford and Don Smith down to Bismarck for the showdown. As Smith recalls the meeting, "Tom sat down with him, and when we left the two Irishmen were friends. And that was the end of it."

A pilot's business is with the wind.
Antoine de Sainte-Exupery
Wind Sand and the Stars

Chapter 6
Get'em in the Blue

In spite of its critics, the Department of aviation made some impressive moves in its first year. It surprised many by winning funding from The Robert B. Campbell Foundation, enough for Odegard to purchase a GAT-1 simulator of a Cessna 150. It would enable students to train indoors in the worst weather. For others, it wasn't so surprising. The chairman of the board of the Campbell Foundation was Fred Orth, a Grand Forks resident who also just happened to be a member of the supportive state Board of Higher Education and someone well known to Tom Clifford.

Odegard also worked hard in those first years to build a relationship with the Federal Aviation Administration. In the fall of 1969, his department hosted a certified flight instructor's workshop that brought 150 pilots to the campus, including a team from the FAA's own training academy in Oklahoma City. Notably in attendance at that workshop was the chief of the Fargo regional FAA office, Lester Severance.

Odegard had a notorious connection to Severance, one that should have worked against him. Many years earlier, when Odegard was a young hellraiser buzzing picnic tables, it was Les Severance who had seen the numbers on the wing of Odegard's plane and lifted his pilot's license.

But in 1969, charmed by Odegard-as-adult and impressed with the way he'd rehabilitated himself, Severance became a huge supporter of the aviation program at UND. His affection led to numerous joint programs and sponsorships over the coming years between UND and the FAA. In fact, not long after the 1969 workshop, the FAA offered to help Odegard develop a program to train air traffic controllers. It made sense from the FAA standpoint. Controllers were in short supply at a time when commercial aviation was showing huge increases in air travel. Safety in the skies was becoming a more important issue than ever.

Yet at the time, the small Grand Forks airport had no control tower. With FAA encouragement, the city bought a portable tower for about $30,000. The FAA staffed it 16 hours a day with a crew chief and a half dozen controllers. They handled air traffic but also trained UND students who were going through Odegard's new Air Traffic Control (ATC) curriculum. While other small cities had portable towers, this was the first time that the FAA had been involved with a private operation. It was an even bigger deal because at the time more than 50 small towns across the country were vying to be first to get federal funding to put in a permanent tower at their local airport. The FAA involvement in Grand Forks essentially put the city at the head of the line, and it got one of the first federally-funded permanent towers.

Two years later, in 1971, Odegard graduated his first three students with an air traffic control certification. It was a particularly gratifying moment for Clifford. He had worried that some students could spend a good bit of money learning to fly and then find out they didn't like it or perhaps be disqualified from flying because of some medical condition. He remembers asking Odegard, "What happens if we get a kid in the program, and his eyes go bad or he has a heart problem?" The answer: that kid was out of luck and a good bit of money.

"So we had to find a way to put that kid in business," says Clifford. "I said, 'We're going to have something for that kid to fall back on. This was very appealing to parents who said, 'Gee, we're investing $20,000 in flight training, what if he doesn't make it?' We had a lot of those kids, and they became very successful airline executives. They didn't fly a whole lot, but they understood the business and had a lot of passion for it. Other schools with aviation programs were not driven by the same idea. They were vocational, we weren't. The big decision for us was to give the person a college education."

To build on the idea of providing business training as well as flight training, Odegard got North Central Airlines—which later became part of Northwest—to send a team led by Gordon Amundsen, the manager of its technical training, to Grand Forks once a week. The team taught students the real world workings of an airline. Northwest Airlines also became enamored of the UND program through one of those back channel connections that Clifford was famous for. It involved Mark Andrews, a Fargo-area farm boy, who represented North Dakota in the U.S. House of Representatives at the time. While in Washington, Andrews had met a fellow North Dakotan named Don Nyrop, then an attorney for the Civilian Aeronautics Board. The CAB was the federal agency that regulated the airline industry until 1970. Northwest Airlines hired Nyrop away from the CAB at about that time, and he quickly became president of the company.

Rep. Andrews used his previous connections with Nyrop to persuade Northwest to increase its service to Fargo and Grand Forks. In the 1960s, Northwest flights coming out of Winnipeg, Canada—two hours north of Grand

Forks—were required to land and clear customs in the small town of Pembina, just over the North Dakota border. From there, the Northwest flights would go on to Minneapolis, bypassing Grand Forks and Fargo. Andrews convinced the Customs agency to set up offices in Grand Forks and Fargo, allowing Northwest to skip Pembina entirely, which it had wanted to do for some time. It could then fly directly to Grand Forks or Fargo for its customs inspection.

A long time friend of Clifford, Andrews hooked him up with Nyrop, and they became fast friends. After that, Northwest officials became frequent guest teachers in Odegard's classrooms. One of the most impressed of those guest executives was Joe Lapensky, a plain-talking Minnesotan who eventually sent his son through the UND program. Lapensky became a passionate believer in the university and would have a major impact on the growth of the aviation program when he later succeeded Nyrop as president of the airline.

Over the years, John Odegard became quite adept at recruiting big names from the aviation and aerospace worlds to his school. Bill Shea, the FAA official Odegard hired in the eighties, describes his brash style. "When I was with the FAA, I'd be at a conference with him, and in the room you'd see all these big shots sitting there. The conference would end, and John would go around shaking hands with everyone. He'd say, 'Great job. Listen you ought to come up to North Dakota. I'm telling you, you would absolutely love it. And there'd be some good finances involved. You'd absolutely love it.'"

More than once Shea watched as the targets of the recruitment reacted in amazement. "As if to say, 'Who the hell is this guy?' That's what he did with me. That's how I met him. After this meeting he comes up to me and says, 'Come up to Grand Forks.' For months we'd meet, and he kept saying, 'Billy we gotta get you up there.' He knew I loved universities. He knew I loved to fly. He'd say, 'We've got a whole bunch of planes up there. Come on Billy we gotta do this.' After four or five months of saying, 'no,' I said, 'okay.' He was a mini-dust devil. He would go to a meeting, and honest to God he'd offer a guy a job."

But as with Don Smith, some of Odegard's best hires—especially in those early days between 1968 and 1974—were people right in his own backyard. Very often they were people who pretty much stumbled across Odegard's program at the very moment they were most needed. If it weren't for Don Smith already serving as the classic example of this type, one would have to award the honor to George Hammond.

He grew up on a farm in Oklahoma and, as a boy, thought flying sounded like fun. He remembers the day, at age 19, when he and his father were hauling a load of fruit and vegetables from Colorado back into the Oklahoma panhandle. They heard on the truck radio that Adolf Hitler's troops had invaded Poland.

They listened for a time to the horrors relayed by the broadcaster. George had completed a semester at what was then Oklahoma A & M, now Oklahoma State University. He was supposed to go back to school in a few days. He turned to his father, who had been a soldier aboard a troop ship heading to France in 1918 when the armistice was signed.

"Well," he said, "I think I'll enlist."

"Yeah," his father said at last. "Go ahead."

Hammond enlisted in the Army Air Corps and was trained as a bombardier. On December 24th, 1941—less than three weeks after Pearl Harbor—Hammond was a tech sergeant in a Martin B-26 bomber flying off the Pacific coast near the state of Washington. He spotted a Japanese submarine coming up out of water. The pilot whirled the plane around, and Hammond got the bomb doors open and let one drop. Because the plane had come around so fast, the bomb missed. But when it came back around again, Hammond dropped several others that scored a direct hit, breaking the submarine in half.

No one had ever successfully bombed a submarine before. It hadn't seemed possible. But now that it had been done, the Army wanted to keep it a top secret. That night when Hammond got back to his base in Tacoma, he wasn't celebrating. Somewhat shaken, he told his wife Alma, "I just killed a bunch of people. You can't tell anyone anything about it." It would be 25 years before the sinking was declassified.

By then, Hammond was a skilled combat pilot, having earned a commission at the end of the war and been accepted into pilot training in the new U.S. Air Force. He served two tours in Vietnam, flying more than 100 successful missions over the North. During his second tour, flying a photo reconnaissance mission over North Vietnam, he says, "I just about bought the farm. They shot nine SAMs [surface-to-air missiles] at us, and we looked at every one of them as we dodged them." For staying the course and completing the mission, he was awarded a Silver Star for heroism.

From Saigon, Hammond was assigned to the Grand Forks Air Force Base as the deputy combat support group commander. He picked up his wife and daughter in Honolulu on the way to North Dakota and remembers saying, "Don't worry honey, we'll be out of there in six months." He turned out to be off by 22-and–a-half years. Here's what happened:

A full colonel during his stint in North Dakota, Hammond had served nearly 40 years in the Air Force, but never made general. The reason: he'd never finished his college degree. During his Grand Forks assignment, when his youngest daughter was a senior at UND, he decided to retire and finish his degree right there.

Thus it was a foregone conclusion he would cross paths with John Odegard and that, naturally, the two would become close flying buddies. Although Hammond had been a flight instructor in the Air Force, Odegard convinced him

It wasn't long before the paths of
Odegard and George Hammond
crossed. He became the director of
flight operations in the early 70s and
ran the airport half of Odegard's
aviation department with the
organization and discipline of a
fighter wing. Even so, the retired
colonel was adored by his students.
He later served as assistant dean and
chair of the department of aviation.

George Hammond,
shown here with his
infant son in the
mid 40's, was the
bombardier on a
Martin B-26 in the
early days of World
War II. His plane
was the first ever to
sink a submarine—
a Japanese U-boat
that surfaced off the
coast of Washington.
Hammond served as
a distinguished
fighter pilot in Korea
and Vietnam before
taking command of
a wing at the Grand
Forks Air Force base
in the early 70s.

to get his civilian instructor rating, which he did. When Hammond finished his degree, Odegard hired him as his chief of the flying program.

It was a decision that didn't go over right away with Jerry Nelson, a retired Air Force sergeant and Odegard's current dispatcher for the airport program. "I remember John telling me, 'Jerome, I'm going to hire a guy who is retired from the Air Force. A colonel.'"

Nelson, whose real name, Gerald, never quite sunk in with Odegard, shuddered at the memory of officers he'd worked for. "I said, 'Oh no, John, you don't need no colonel.'"

Nelson ended up working directly for Hammond but was quickly impressed that this was no ordinary oh-six. "What a fine gentleman he was," says Nelson. "He and John were terrific together. They had the same personality."

From Hammond's standpoint, Odegard's personality went like this: "He was a real tiger. A goer. John had lots of ideas many of us thought were impossible, and he made them happen anyway. He didn't look at anything as something that was not possible. He didn't like no. He was a real performer."

Al Palmer, a man Hammond hired more than 30 years ago—and who now runs flight operations just as his mentor once did—compares the two this way: "George Hammond was five foot two, and John was very tall. But George was as enthusiastic as John was. He was the first one out here in morning and the last one to go home at night. He would walk around with a smile on his face, and he'd have a positive attitude that just translated right on down. We just all picked up on that."

Palmer was one of a trio of young men—along with Dana Siewert and Don Dubuque—who were hired by Hammond in the seventies. All three men lacked a college degree. Mindful of what the lack of a degree had cost him, Hammond hounded the trio mercilessly to go back and finish their degrees. All three eventually did. Today, Dubuque is the director of extension programs; Siewert, a former head of flight ops is now the award-winning director of safety. Palmer is not only the head of flight ops, but he now outranks his old mentor.

Palmer had been an enlisted man at the Grand Forks Air Force base when he decided to learn to fly. Curiously he couldn't do that in the Air Force. So he got his certifications through the UND program. Hammond then hired Palmer as a flight instructor. For several years Palmer taught flying to Odegard students by day and then worked night shifts for the Air Force. He left the Air Force after nine years to become a full-time employee of Hammond's. The old colonel immediately brow beat Palmer into going to college part time.

"He just pushed it and pushed it," recalls Palmer. "It's one thing to say you need to do it, but he also gave us tools. We had to go to school in the daytime. He said take time off if you need to it in daytime, but I expect you to work on

To Odegard's great fortune, his benefactor at UND, Tom Clifford, Vice President for Finance, became university president in 1972. For 20 years as president, Clifford supported Odegard's expansive dreams of a self-contained aerospace college. Clifford continually deflected criticism from a jealous faculty, resentful of Odegard's lack of academic standing.

As Odegard was mentored by Don Smith and Tom Clifford, he himself became mentor to a handful of young men only a few years younger. One of them, Alan Palmer—shown here in the early 90s—hired on as a part-time flight instructor while still an airman at the Grand Forks Air Force base. Today he is UND's Director of Flight Operations and a brigadier general in the North Dakota Air National Guard.

weekends and at night. And boy we put in our 50 to 60 hours a week because we didn't want to disappoint him. But we also wanted that degree."

Hammond also pushed Palmer to join the Air National Guard unit in Fargo because he already had nine years toward a military pension. Thus, Palmer went to college part time and to the National Guard on weekends, all while holding a full time job. His commander in Fargo got him into Air Force officer candidate school and Palmer began slowly working his way up the ranks. In the spring of 2006, 20 years after being commissioned a second lieutenant, Palmer was promoted to Brigadier General in the Guard. The first person he called was George Hammond.

"George was very good at motivating people," says Don Dubuque. "He was a wonderful guy, very dynamic."

Hammond was well known at the airport for his daily mantra to instructors: "Get 'em in the blue." Until his arrival, operations at the airport were somewhat disorganized. Students would either all show up at the same time to take their flight instruction, or long hours would go by when nobody showed up. Not only did that leave aircraft idle, it put a serious crimp in the only cash flow the school had— payment of fees by students for actual flying.

Hammond took over and ran the operation exactly like he ran the fighter wing at Grand Forks Air Force base. The mission, as he saw it, was to get the maximum use out of every plane. That meant getting students into those planes and making sure they earned all the hours they needed to earn their various flight certifications—and to keep the program funded.

A man who appreciated that efficiency was Bob Reis, Odegard's first director of fiscal affairs. "George was a 'get em in the blue' guy," says Reis. "Those were the first words out his mouth everyday. If they were awake and able to fly then 'get 'em in the blue.' It wasn't whenever you had time. You were now scheduled. You will be at the airport at this time, and we are going to have planes on the ramp and ready to fly. Everything had to be oiled and ready to go in motion."

Quality-wise, says Reis, Hammond knew that volume of use was a strategy that allowed the department to buy the best planes. "If you get more volume, you can bring your costs down to make it more affordable to the student customer," says Reis. "George was a great motivator."

In spite of his firm hand, Hammond was popular with his students.

"George loved young people," says Don Smith, "and he was loved by the students. He put the operation of the airport on a sound management and flight footing, in terms of discipline, in a nice way."

Both George and his wife Alma became such favorites of students that they were often invited to student-only parties.

"I think the greatest good for me was getting to know that younger crowd," says Hammond, now 87. "They treated us as one of them, and we were probably older than their parents."

With Hammond and Smith now part of the team, Odegard had two able lieutenants he could rely on. In many ways, they were just like him. Somebody back in those days took a photograph of Odegard, Smith and Hammond together. It was posted in the Gamble Hall office without a caption. But it quickly became known as "The Holy Trinity."

Motor cut. Forced landing.
Hit cow. Cow died. Scared me.

Dean Smith, telegraph to his chief,
as quoted by Amelia Earhart
'The Fun of It,' 1932

Chapter 7

The Bunkerdoodle Effect

In Rushford, Minnesota, a small town two and half hours southeast of
Minneapolis, they're still shaking their heads over the day in 1977 when John
Odegard popped up out of nowhere.

Rushford, population about 1,300, is where Jim Bunke grew up. His father,
Rob Bunke, was a businessman who helped put together small town telephone
companies in the region. His first love, however, was airplanes, and he'd bought a
Piper Cherokee 140 to fly to places like Des Moines, Iowa, or Madison Wisconsin
or up to St. Paul for business.

He often took his son Jim along, fueling a boyhood love of flying that has never
quit. Jim soloed on his sixteenth birthday, got his private license at 17 and his
commercial two weeks after his eighteenth birthday. By the time he arrived at the
University of North Dakota as a sophomore transfer from Indiana State, he already
had a flight instructor's certificate. George Hammond was happy to hire him as a
part-time instructor.

Bunke, an ambitious, effervescent young man with an engaging personality, was
an immediate hit with his fellow students in the department of aviation. He was
elected president of the local chapter of Alpha Eta Rho, the aviation fraternity
which gave him a seat on the student aviation advisory council. It reported to the
chair of the aviation department, John Odegard.

Bunke's status and easy sense of humor often landed him the job of emcee at
banquets hosted by the fraternity or the department. One of his duties was to
introduce Odegard at the end of a dinner. Bunke would usually poke fun at
Odegard for his classic habit at dinner meetings of introducing practically every
person in the room and always having something nice to say about each.

There were lots of those dinners and banquets which always seemed to run late.

"But they gave me a chance to bond with John," says Bunke. "I addressed him as John, and he addressed me as Bunkerdoodle.'"

Because the department was so small and intimate in those days, and because of all the bonding Bunke had done, he felt emboldened enough to ask Odegard an unusual favor. Bunke's father, the aviation buff, had pushed his fellow citizens back in Rushford to build a community airport. He'd helped develop an airport authority which lined up enough funding to build a 3,400-foot grass airstrip to be known ever after as the Rushford Municipal Airport. In 1977, it was time for a dedication ceremony, and Bunke's father was hoping to stage some kind of air show.

When Bunke asked Odegard if he'd do the show, he was surprised at how quickly his mentor agreed. Odegard had previously wangled the donation of a free glider from a manufacturer, and he told Bunke he'd put on a one-man glider air show.

Odegard flew into Rushford in a Beechcraft Bonanza, while several student volunteers drove the whole distance from Grand Forks towing the partially disassembled glider. As Odegard flew into Rushford, he carefully surveyed the terrain. The town was situated in a very hilly, almost mountainous part of the state. The town was 500 feet above sea level, but the airport sat on a hill about 700 feet higher. The area was dotted with steep hills and cliffs.

The day of the show, excited neighbors flew or drove in from three surrounding counties. As a tow plane took off pulling Odegard into the air in the glider, a master of ceremonies on the ground named Sherm Booen kept up a running commentary for the crowd. Everyone knew Booen. He produced and hosted a weekly TV show on WCCO-TV out of Minneapolis entitled "The World of Aviation" and also published a monthly magazine called "The Minnesota Flyer." To everyone at the air show that day, he was known simply as "Mr. Aviation."

As Booen addressed the small crowd, music from "Jonathan Livingston Seagull" –Odegard's choice—played over the loudspeakers. High above, the tow plane released the glider, and the show began.

"The music is playing," recalls Bunke, "and people are just in awe. The glider has these big silver wings. It just showed up beautifully against the sky. He was doing incredible aerobatics with it. And it was all silent."

Bunke's father still remembers the day. "There wasn't a sound from the entire crowd," says Rob Bunke, "only the beautiful music in the background. I have often said that it seemed like John could hear the music, and he flew as if he was directing the song. I still get the chills thinking of that beautiful performance."

Odegard went through his entire routine and started his final maneuvers. All he had told Booen was that the show would end with a spin of 21 revolutions. Booen announced this to the audience, and as Odegard began his dive everyone began counting aloud. "One…two…three.."

In his pre-show survey of the terrain, Odegard had seen exactly how to plot the final revolution for maximum effect. In fact, Rob Bunke was acting as an ex-officio safety officer that day and had talked about the stunt with Odegard beforehand, satisfying himself that it was doable.

Mr. Aviation and the assembly of Minnesota country folk continued to count off the revolutions as Odegard and his glider spun ever closer to the ground. When they had reached a count of 18, Odegard's glider suddenly dipped below the horizon line. It disappeared completely from the view of the spectators at the far end of the hill on which the airport sat.

Mr. Aviation stopped counting. The crowd started buzzing. The glider was nowhere to be seen. Sherm Booen began frantically trying to calm the crowd down. "We know John is a professional," he stumbled, "and even though we can't see him, we're sure he has landed safely in a hayfield down there."

But everyone knew what had happened. The glider had crashed and likely killed its pilot.

"Well folks," Booen fumbled again, "there are lots of other things to do today…"

Then, after another few seconds of dead air he grew ever more serious. "We're certain the authorities will let us know, as soon as they can, that they've found John…"

At that high moment of tension, recalls Bunke, "Poof! John comes shooting up from behind this bluff from 700 feet down in the valley. He lands the glider, and the crowd is screaming in delight. What a spectacular ending to the show."

But it wasn't over yet. When Odegard brought the glider down to earth, he rolled directly toward the crowd, veering off at the last minute to a final stop. Bunke's father rushed out to the runway to greet him. But he was followed by an irate man from the crowd, a man screaming about the unsafe maneuver. The two men reached the glider at the same time, as Odegard was climbing out. While Rob Bunke was congratulating Odegard, the man from the crowd identified himself loudly as an official of the FAA. He berated Odegard up and down for his disappearing act, then landing in a path directly toward the audience. "You could have wiped out dozens!" he shouted.

Odegard remained calm, almost amused. He walked the FAA official to the edge of the hill and pointed out in detail his entire flight plan, including a man-made pond down in the valley that was to serve as his escape landing area. He turned and indicated where the crowd was seated and how he had calculated his landing speed and direction to the nth degree. But the FAA official wasn't having any of it and stomped off to file a complaint at the district office.

Bunkerdoodle, meanwhile, was a mass of anxiety. "I'd asked John to come all this way, and he does it out of the goodness of his heart, and he's the chair of the program, and now he's about to lose his license."

In fact, the complaint was filed but later thrown out, due to Odegard's vigorous defense of that detailed flight plan—plus an endorsement of its safety from Rob

John Odegard's charisma and dedication influenced the careers of scores of his students, including that of Jim Bunke, an impish student whom he called "Bunkderdoodle." Odegard convinced Bunke he'd make a great salesman and helped him win his first job at Beechcraft. Shown here with his mentor, Bunke went on to become national sales director of Bombardier business aircraft in Minneapolis.

In the early 70s, when his aviation program was new and relatively small, Odegard was able to stay close to almost all of his students. Here, during a field trip to European airshows, Odegard and Diane (far left) and nine aviation students hoist a mug in a German beer garden.

Bunke. At the moment, though, not knowing what would happen, Odegard wore his usual smile and exhibited his usual charm as he shook the hand of Bunkerdoodle Sr.

"I see what you mean," Rob Bunke later told his son. "This guy is smooth."

The glider incident deconstructs Odegard down to his basic iron: a passion for flying, a willingness to take risks—although, as he pointed out to the FAA official, *perfectly* calculated risks—a knack for getting into trouble, a balancing ability for getting out of it, a fervor for planning something complicated so it appears both breathtaking and easy. And, of course, an affinity for helping out a student.

For in spite of all of the grand personal achievements during his life, Odegard's deep-felt need to give some intangible piece of himself to his students—done usually in quiet moments without fanfare—stands as a more meaningful measure of his character. The English novelist E. M. Forster once coined a term for fictional characters whose one-note droning existence didn't ring true to life. He had in mind the Dickensian panoply of Scrooges and Tiny Tims whose exaggerated black and white shadings spoke of a cartoonish reality. Forster called them flat characters—as opposed to those rounded people in stories whose multiple, often conflicting traits not only make them more interesting, but more human and therefore more credible down the ages. For Odegard, a man who in almost every phase of his life never stopped trying to show off to the world a perfectly executed flight plan, ministering to the needs of his vulnerable Bunkerdoodles gently rounded him into three-dimensional imperfection.

"I would say his drive was his students," says Pam Kvidt, Odegard's assistant for a decade. "And yet he always felt he was disappointing them."

Close friends of Odegard believe both his drive and fear of disappointing came from a vague yearning to prove something to his father with whom he was never as close as with his mother. Brothers John and Jim often were sent out by Clara to collect Truman from a saloon. Others say John felt a desperate need to build a fuller life than his parents had. But the guilt that Kvidt describes may be traceable to Odegard's own poor performance as a high school student. His friend Del Rae Meier recalls his frequent reluctance through his adult life to discuss the old Minot high school days. "He felt badly that he didn't make enough of those years," she says. "He used to say, 'Oh, I only played in the band.' I think he was disappointed in himself."

He made up for it in several ways, some more rounded than others: there was the typically grandiose Odegard fashion, visible to all. But there were also countless subtle gestures without benefit of audience and often having little to do with him.

In his grand manner, Odegard seldom missed a student banquet; not only did he allow his students to make fun of him, or roast him, he often used such events as an opportunity to make sure they all got some kind of recognition. For example, he insisted that suppliers and vendors who wanted to do business with him contribute money toward scholarships. Russ Watson, who attended the annual scholarship dinner at UND to present the Cessna scholarship, remembers that every year virtually every student received some award.

"I never ever heard him say 'I,'" says Kvidt. Indeed, it was an Odegard trademark to always use the royal 'we' in his correspondence. "He never took the credit. He was really very humble. At any banquet or event everybody would say good things about him. But then he'd get up and give credit to Tom Clifford or Don Smith or his staff. He inspired people by that."

Certainly his public reputation as a selfless, even-tempered leader grew from such episodes. "He had a personality you wouldn't believe," says Jerry Nelson, the former sergeant turned dispatcher. "That guy put in a lot of hours, from 6 a.m. to 8 p.m. many days. He seemed bubbly all the time. In 23 years I never heard anyone say a bad word about him."

If this were fiction, Forster would wave his flag of flatness. In fact, there were many who had less than complimentary words to say about John Odegard, including, on occasion, his protector and father-figure Tom Clifford.

"John was very affable, outgoing, vigorous," says Clifford. "He was a hard worker, very hard. People forget that. He was also quite arrogant. He'd go and park where he wasn't supposed to. Just drive up. I'd tell him not to do this. He was not aware of the fact that sometimes people don't like that. You have to work with that a little bit. And we did, and he did. He had to modify his behavior. He treated some people well, but he treated some people poorly. He made a lot of people angry because of that."

Uncomfortable contradictions, such as humility and arrogance, play hell with hero worship. But in real life, Odegard labored under the same handicap that burdens everyone: the quirky human condition. An outsider hearing of his stunning exploits from afar might wonder how anyone could be so perfect. The curse or blessing of roundness, however, guarantees that no one's sainthood stands up to the scrutiny of those devilish details revealed by a magnifying glass.

One has only to look at the arena of modern sports, with its almost daily regurgitation of off-field blunderings by otherwise perfect home run heroes and gridiron legends. The endless feet-of-bronze versus feet-of-clay comparison in the ephemeral world of sports may seem at times absurd. But in real life such scrutiny is not an insignificant side trip on the road to understanding why and how something of lasting social value is often founded on a mixture of humility and arrogance. Without his constant drive for perfection, whatever the psycho-social *Ur text*, John Odegard might have ended up an unhappy accountant in Minneapolis. With it, he made things happen in moments ranging from the brilliant to the banal to the hilarious—sometimes all at the same time.

Consider the bedrock philosophy of Odegard's life, a home-spun proverb he preached ceaselessly: "You never get a second chance to make a good first impression."

He meant it on several levels. First, of course, from the all-telling moment of the introductory handshake—Strong like a bull? Or limp like a fish? He believed that strength or weakness revealed itself in the eyes and in the personal affect: a weak handshake, a restrained greeting or a reluctance to meet a man's eyes said more to him than any words about one's likelihood of achieving success.

But Odegard also placed enormous emphasis on one's attire.

"He loved clothes," says Jerry Murray. "John was the best-dressed man on campus. He always looked the part."

An examination of old yearbook pictures of the pre-Odegard flying club shows members dressed in standard, college-student garb: the rumpled, the ratty, the remaindered. Once Odegard stages his coup, the photos begin to show evidence of revolution: members wearing trendy polyester suits. Later, as the plastic embarrassment of polyester became apparent, club members appear in the neat, well-groomed apparel of a professional.

"He was always a clean dresser, but no more than average," says Odegard's brother Jim. Soon after John's marriage, though, he stopped buying shirts and suits off the racks. Whenever he was in Minneapolis, he would make a raid on a Nordstrom's or a Brooks Brothers. He shopped quickly, never debating over a choice, and seldom spending extravagantly. (Later, when his department had grown into its own school, he ordered hand-made suits and monogrammed shirts from Taiwan.)

Unlike many fashion plates with a closet full of clothes that never get worn, Odegard viewed his wardrobe as a tool to be used. If he bought it, he wore it.

"It was important for him to appear on top of it," says Jim Odegard. "Part of it was 'If I work with legislators and businessmen, I want them to think they are dealing on solid ground with me.' And I think he gave that impression."

In Odegard's department, impression was a commandment. There were never dress-down days. One thing guaranteed to trigger his outburst was to see a faculty member wearing jeans or even shorts. If he could have—and many times cooler heads talked him out of it—he would have established a dress code. As it was, his conservative favorites—a camel hair topcoat, Burberry scarf, shirt, tie, blue blazer and tan gabardine slacks—became the unofficial uniform of those men who worked for him and counted themselves as savvy.

"He was very professional looking," says Diane Odegard. "But he was such a perfectionist." So much so, that he insisted on ironing his own shirts. "I was pretty good at ironing, but no one could do the shirts like he did. If his shirts went to the cleaners, he would re-iron every one when they came back. He didn't feel they could get the cuffs and collars right. I thought, 'God, this is ridiculous. I'm not going to do this.' But sometimes I did."

The ironing compulsion spread to other areas of his home life. Odegard, like his father and Uncle Al, was a stickler for detail on Harry Homeowner projects. His was a well-known face at the local hardware store in Grand Forks where he seldom left without buying half a dozen of any gizmo he needed. He never lost his boyhood fascination for building things, from furniture, to a dock at his lakeside cabin, to painting the house.

"He always finished projects to perfection," says his daughter Stephanie. "He'd even clean up the sawdust. It's not the way I am. I'd be saying, 'Let's hurry up,' and he'd say, 'No, we're going to do this right.'" Even after Stephanie left home and started her own career, her father couldn't pay a visit without finding something in her apartment in need of improvement. "He'd look at the pictures on the wall and say, 'Everyone hangs art work too high. They're supposed to be hung at eye level.' He'd go off to the hardware store and come back with new hooks and re-hang everything."

"He always seemed to do what needed to be done," adds Diane. "It wasn't like he was sitting on the floor playing games with the kids for hours and hours. He never changed diapers or fed the baby or anything that men do now. He didn't cook. He didn't clean up the house. I did those things. I never expected him to do those things. But he did fix things, repair things, made sure we had things. We tried wallpapering together but that didn't work. I was mad at him, and he was mad at me. But he was a great painter. His big thing: get enough paint on the brush, put it on thick and smooth it out. He learned it from his father."

John never had disparaging words to say about Truman, but it was clear that he wanted more from life than his father's sedentary, beer-fueled jolliness or moodiness.

"He wanted more. I knew that right from the start," says Diane. "His was not a poor family, but not wealthy either. Not even middle class. Just a working-class family. They did not have very much money. He didn't want that kind of life. He always wanted things to be better than they were. He wanted to paint his parent's house to make it look better. But he wanted the job done right. His attitude: You never give up until its done right."

Diane remembers the time they brought in a contractor to remodel their kitchen. Before he went to bed that night, John seemed vaguely unhappy about the job but couldn't put his finger on why.

"He woke up in the middle of the night," says Diane, "and said, 'Goddamn, I know what's wrong with that kitchen. They put the wrong size hinges on the cabinets.'

"He was right too. It was a little thing but they had installed the wrong size brass fixtures. He woke up, and he called the people up, and he got it redone."

The airplane is a means of getting away
from towns and their bookkeeping
and coming to grips with reality

Antoine de Ste-Exupery
Wind, Sand and the Stars

Chapter 8
Things That Carried Us Through

John Odegard's impatient perfectionism went totally against what he perceived as the work ethic of the faculty across campus. "This was not an 8 to 4:30 operation," says Bob Reis. "John found people who worked whatever hours it took to get the job done. I remember telling a staff member, 'I need you to work this project out.' This poor gal worked so hard. Her son was having a birthday party that night. I remember her crying on the phone. But she says, 'I'll do it. I'll get it done.' She came in and worked on it and got it done. That's the type of motivation he developed in all of us. Wasn't always easy. We had moments of stress. We pushed our people to the limits."

"It was his weakness," says Pam Kvidt. "He set his standards sometimes so high he was disappointed if they weren't met. People didn't want to disappoint him. If you did, you would know it. He would tell you. You would know."

Odegard was supremely organized but often to his own peculiar standard. He kept three calendars, the one on his desktop, the one Pam Kvidt had on her desk and a pocket calendar he took with him. Keeping track and coordinating them was close to impossible; his schedule changed almost by the minute every day.

"No one on staff ever made appointments with him," says Kvidt. "They knew even if they had an appointment they would have to wait. I remember how frustrating it would be when he did have an appointment, because somebody would shoot in there while somebody else was waiting. And he didn't like to cut his meetings short whether scheduled or unscheduled. When he wanted to connect, he would devote his full attention to you. He would never let anyone know he was rushed."

Yet Odegard usually arrived at work in a rush. He spent most of his time either in meetings or on the phone. In many of those meetings, he harped on the same theme.

"I remember being in a meeting where someone would say, 'I'm not sure we can do that,'" recalls Dana Siewert. "John's response was always, 'I don't want to hear how you can't do something. I want to hear how you can do it.' We always figured out a way to get something done, although sometimes it was pretty long hours."

Most of those who relate such stories do so with a mixture of bemusement or a measure of pride at having survived.

"No one thought he was an ogre," says Terri Clark, today's director of fiscal affairs. "There were times we got frustrated with him. Probably more out of exhaustion than anything else. We were tired trying to keep up, and you couldn't keep up with him. I don't think he sensed that."

"You had to have energy," adds Don Dubuque. "You couldn't just sit around and maintain the status quo. The people who did that didn't last long."

Even those who had energy, says Kent Lovelace, often ran out of it and went to the boss to complain. "You could go into him with a problem, ready to quit, and you'd come out with three more jobs, no more pay, and feel on top of the world. He was very commanding, very positive and very charismatic."

One year, the annual NIFA air competition was supposed to be held at Alabama, but a hurricane forced cancellation. Seeing an opportunity, Odegard volunteered UND on short notice to host the prestigious event. The immediate complaints from his staff: too little time to prepare and zero money in the budget.

Lovelace remembers reporting weekly to Odegard on the status of the planning. "He'd say, 'Well let's do this.' I'd say, 'We ain't got the money.' He'd say, "We'll worry about that later. I want this to be something people will remember.'"

To accommodate the influx of all the aircraft expected, Lovelace had crews laying tar to expand the runway ramp at the airport just four days before the event.

"I can remember standing out there with John," says Lovelace. "He said, 'Now don't worry Kent, it's going to be fine. It'll work out.' He was that kind of guy. He never let obstacles get in his way. There's a few of us who joked that we were like circus workers, because we were always following the elephant, sweeping up and trying to keep up."

Ultimately, it was hard to complain about long hours, says Lovelace, because the boss himself worked like a dog. "You could come in here on a Saturday, and he'd be in his office working. I don't know if the job was his hobby, but it was a true love and true passion. He always said to us that being a faculty member, being in a salaried position doesn't mean you work 40 hours a week. If you're not working 50, you're not working hard enough. He put enough energy into it, so you had to do the same yourself. It was leadership by example."

As a man always concerned about "the look," Odegard worked hard at cultivating the impression that those long hours were almost a hobby. Whenever someone commented about his drive, he would shrug it off and quote the laid-back wisdom of Confucius: "I've never worked a day in my life."

Yet as sure as the sky being above ground, Odegard was never close to being laid-back. Diane saw that at home in a variety of ways, including the way he prepared himself for speaking engagements.

"He had to give a lot of speeches, and he constantly worked on them. He was not like Tom Clifford who never had any notes and could speak anywhere anytime. John could get very uptight over how a speech would go, yet he wasn't ever going to read anything. He was going to get up and talk like he never had a worry about it in the world. That's how he came off. When he gave a speech he looked like 'Wow how could this guy just get up and do this?' But he *did* worry about it, about all the details. I knew, because I was here with him. He worked so hard all the time. And there was always something at work that could be a major problem. We took those to heart here at home. Always something."

One night in the early seventies, the Odegards were watching television together in the living room. Something came on about a concentration camp, something that suddenly struck John as unbearably awful.

"I feel like I can't breathe," he told Diane.

She remembers him getting up and saying he had to go for a walk. "We walked and walked. He said, 'I just don't feel right.' He was very scared. It was during those early days, that very tenuous time when nothing had totally started yet. We had students and things going on, but it was like what are we going to come up with next? There was no way to know how the work would go."

As he often did, John sought out his older brother for advice. "At one point I was worried a little bit," says Jim Odegard. "The school was growing by leaps and bounds. We happened to be talking, and he said, 'I can't sleep at night. I lay there and think of all that's going on and how I'm going to control it. What kind of a monster have I created here?'"

Diane remembers John insisting to her that whatever was bothering him wasn't depression. "If I ever got depressed," he'd say, "I would just go flying." Soon after, through a racquetball partner, Odegard got the name of Dr. Harold Randall, a psychologist at the university's rehabilitation center.

"He knew he had to do something," says Diane. "There were things he had to face up to that were bothering him, that he was worried about. He was a sentimental person. He always said that when he watched Spartacus, at the end he felt like crying. He had that soft spot. Things touched him. Anything to do with children. He was very protective of them."

Today, Diane believes John was typically forward thinking in his decision to seek professional help. At the time, though, she was troubled.

"Seeing a psychologist? In Grand Forks? To me, as a young mother and wife, I didn't know anything about people going to talk over problems with therapists. I

Odegard often combined business trips promoting his aviation and aerospace programs with pleasure. An avid outdoorsman, Odegard loved to ski and was a frequent visitor to the Colorado slopes. Here, Diane (second from left) and John (far right) pause for a ski slope portrait with close friends Rob and Judy Larson.

With a plane always at his disposal, Odegard often took his entire family with him on ski trips. Top left is John, Jr., standing next to his father. At lower left, Stephanie stands beside her mother.

was pretty upset about this and about how he'd come out of this. I was very fearful. I remember standing at the sink and thinking he's just going to have to decide to get better. My way of thinking about it was you just get out there and do it. It's like telling a person don't be depressed. He's going to have to decide he just doesn't want to have anxiety attacks. It was very, very not helpful thinking. It wasn't very modern."

Randall talked to John about the need for relaxation. Not many people who knew Odegard can remember even a handful of times when he seemed relaxed. "But Harold was very proactive in mind-body connections," says Diane. Part of Odegard's therapy was to write long notes to himself. Diane remembers him filling up several yellow legal pads during this period. After about four months, Odegard's anxiety and panic attacks passed. "I can't tell how it passed," says Diane. "It just worked itself out. He got through it just fine, and he never had that happen again."

To stave off future trouble, though, Odegard tried vitamins and herbal homeopathic remedies. If something new came along, he tried it. He even got into the fad of the moment, Transcendental Meditation.

"When TM came along it was a pretty weird thing," says Diane. "A teacher came up from Fargo and had a meeting in the student union. You got a mantra, a secret word. John kind of got into this. We had a close friend, Judy Larson, who was having blood pressure problems. John said, 'I think Judy could benefit from this.' Judy's husband Rob was the type who would do anything she did.

"So pretty soon the three of them are going to these meetings, and they had these secret words, and I thought, 'Wait a minute. I'm kind of left out of this circle.' I was kind of jealous and feeling threatened. What are they doing there? I don't have a secret word. He'd come home, and he'd say he had to meditate for 30 minutes a night. He'd go upstairs and get into his black recliner. The children couldn't say a word. And I knew he was up there saying his secret word, which I didn't know. So I got completely bent out of shape."

At the time, the Larsons and Odegards were planning a Christmas ski trip together. During the planning, John announced that he and Judy and Rob would need to set aside time to meditate. Diane was furious.

"I thought, 'Even on the ski trip!' So I started reading books about TM. I was fired up, because I felt left out. He was doing it to help himself to relax. But I was taking it as a personal affront. I practically worked myself up into a nervous breakdown over this."

Eventually, Diane enrolled in the local TM program. One of the requirements of new members was attendance at a special ceremony. "I had to bring flowers and fruit and all this stupid stuff," she remembers. "It was held at the Ramada Inn. And the three of them had already done this. I said, 'I'm going to go to the meetings, and I'm going to get this secret word.' And I did, and then I started meditating too."

Diane tells a hilarious story of how John came back from a TM mediation session only to find his favorite ski boots had been ruined. His disappointment was legendary. (Here he demonstrates his form with his famous yellow ski boots, shown here in living black and white).

In the early days of the program, Odegard suffered a bout of anxiety and sought professional help. He began relaxation exercises and even joined the Transcendental Meditation movement for a time. Skiing with Diane and friends was a favorite Odegard method of unwinding.

During the ski trip, though, the TM triggered one of those memorable incidents that end up as legendary stories repeated at family gatherings for decades. One day, after a particularly wet snow on the slopes, the four friends came back to their quarters and each found a separate room to do their meditating. John left his ski boots on top of a heater vent to dry.

"These were his beautiful yellow Nordica ski boots," says Diane. "He talked about these boots all the time. But when we came back from meditating, we saw that the boots had melted from the heat of the vent. I said, 'Oh, John's boots.' And John went 'Goddamn it, the boots are melted.' He started trying to get them back in shape. Judy and Rob started laughing. I had tears in my eyes. The more upset John got, the funnier it seemed to us, and he really got upset."

Not long after that, the four friends went to dinner and the subject of conversation turned to their TM mantras—the sacred secret word. Diane asked the others what their secret word was. Each was reluctant to divulge it until Rob suggested that it would be all right if everyone just hummed their mantras.

So Rob and John started, each humming something that sounded like "Shareeeeemmmmmm."

Diane's jaw dropped. "Well that's the exact same word I have," she huffed.

Judy then admitted it was the same word she had as well.

"So we realized," says Diane, "that we all had the same secret word—which probably everyone got. It was like a Woody Allen movie. We practically fell out of our chairs laughing and thinking how stupid this was."

By the way, the day after the melting ski boot incident, John called Dave Vaaler to see if his homeowners insurance covered the damage. It didn't, but by then John seemed calm, almost Confucian.

"He could get upset about things," says Diane, "but nothing that lasted very long. He was not an angry man. With children and students and the people who worked for him, he was a great motivator. He had a great ability to say, 'I love you' very easily and show love to the children. Whereas I could get much more upset. I got worried and bent out of shape over things. I was much more of an uptight worrier than he was."

Odegard ended up buying a new pair of ski boots—they were red—and they quickly became the next best, most wonderful boots in the world. Diane instantly got over her jealousy.

"These were the things," she says, "that carried us through."

The whole art of teaching is only
the art of awakening the natural curiosity of young minds
for the purpose of satisfying it afterwards.
Anatole France

Chapter 9
You've Come to the Right Place

In his cramped outpost in Gamble Hall, meanwhile, creating and maintaining a look was less important to Odegard than forging and developing the right stuff in his students. Not that he didn't insist on them dressing as well as they could or making good impressions on behalf of the university. He did that, but more often than not it was to the student's future advantage he was looking rather than an immediate gain for his program.

Jerry Murray says the enthusiasm Odegard showed everyday in the first half of the seventies reflected how much he wanted everyone to know how wonderful an experience flying could be. "We all worked hard everyday to learn to fly," he says. "Pretty soon you were consumed with the same passion about flying."

Odegard's relationship to his students was, naturally, non-traditional.

"There was never this teacher-student type relationship," says Don Johnston. "He didn't want any barriers, so you felt like a colleague. He knew everyone and empowered everyone."

At Christmastime, John and Diane would host "Tom and Jerry" parties for students, and a couple times a year they'd invite all the flight instructors over to Reeves Drive for a social hour.

"It felt like we had a family," says Bunke. "John would come to our frat meetings and have fun drinking with us." Odegard, however, drank scotch while everyone else drank beer.

"John felt a need to take care of us," says Jean Haley Harper—Jean Haley in those days. "He knew we didn't have money. Once he found out about a scholarship and filled it out on his own—one for me and one for Tracy Vandenberg. And we both won it. He stopped by one day and said, 'Congratulations, you won a $400 scholarship.' We didn't know what he was talking about."

As chummy as he could be, Odegard never forgot his own wasted school days. One day, says Haley, she stopped by his office for a chat but his secretary said, "You don't want to go in there—he's with a student."

It was a kid whose father had asked Odegard to take special care of his boy. Even though his door was shut, Haley could hear him shouting, "You got an F in the only class you didn't drop! You signed up for six classes and dropped five of them to go fly, and then you flunked the other! What am I going to tell your father?"

Haley says this first glimpse of Odegard the disciplinarian was eye-opening. Though not quite eye-opening enough. "I saw then 'Don't get on his bad side. Don't mess up.' But I did. One time I got a bad grade in a class. I didn't show up to the class very often, because I had this great part-time flying job. John had a little concerned talk with me. I felt as if I'd let him down terribly. He was more disciplined than I was at that age."

Haley, like Bunke and Murray and many others, felt a special relationship with Odegard, going back to the day she arrived on campus to begin her studies.

Born Jean Haley in rural California near the San Joaquin Valley, she can trace her love of flying to her father, Frank. A crop duster, he was also the fixed base operator in Tracy, California where he ran a flying school. His wife was a nurse who rode a motorcycle, so it wasn't unusual for Jean to see herself as a pilot someday. But in the 1950s, the idea of a girl becoming a pilot was ridiculous enough that Jean's third grade teacher gave her an F and a stinging lecture when she wrote about her dream in an essay on what she wanted to be when she grew up.

She was told to be "a nurse or a mommy because that's what girls do." It was a foretaste of things to come. "I got a lot of garbage in high school about flying, to the point I just closed down. I learned to shut my mouth about what I wanted. I'd tell friends I wanted to be a stewardess. At least I didn't catch a lot of grief."

She started flying lessons at 16 in the mid-sixties. By 1971, she had earned her pilot's license and was certified as a flight instructor. The young FAA official who certified her turned out to be Mike Zachary, a recent aviation graduate of the University of North Dakota. He encouraged Jean to enroll there and smoothed the path by telling John Odegard about her. Odegard not only accepted her application as a student at the university but hired her as a part-time flight instructor.

The early 1970s was a stormy time for women who hoped to find a way into realms that forever had been jealously restricted to men. Women were just beginning to challenge notions that only men could handle certain jobs. There still were no women pilots at airlines, and at UND, there were no women flight instructors. To Odegard, gender considerations seemed completely irrelevant.

Haley had never met him, however, and wasn't sure what his or the school's real attitude was toward women as pilots. In late summer of 1971, when Haley got off

Jean Haley (later Jean Haley Harper), the daughter of a crop duster, became the first female flight instructor at UND while still a student. John Odegard had to read the riot act to some of his airport crew who objected to a female presence. Haley later became the first female captain at United Airlines.

the plane in Grand Forks for the first time, she found a young man with a Volkswagen Beetle waiting to take her to campus.

"I thought 'How nice, Mr. Odegard has sent one of the students to get me.'"

When the young man shook hands with her and said, "I'm John Odegard," she did a double take. "I just wasn't expecting someone that young. He was kind of baby faced. I found out later he was 29. But he was nicely dressed in a three piece suit."

As they drove into town, Odegard told Haley that the program had 50 students, five planes and two more on the way. Then he asked her a question she'd been dreading.

"He asked me what I wanted to do for my life's work. I said I want to be an airline pilot. There was a silence that seemed like an eternity. My future depended on what he was going to say. And I wondered if he was going to snow me and say, 'Oh of course you're going to make it, I guarantee it.' His credibility would have been splattered. If he'd given me some bullshit line, I probably would have gotten on the airplane the next day and gone home. On the other hand, if he had said a lot of the things I'd heard before, like 'That will never happen…' I was so afraid. Finally he said, 'Well, you've come to the right place.' Just a statement of truth. No big promises. No blocks in my path. He showed no shock at my gender or aspirations."

In California, Haley had done a lot skydiving. So the day after she arrived in Grand Forks—a day forever etched into her memory: August 30, 1971—after she'd registered for classes, she looked up a local skydivers club. She made three jumps that afternoon, the only woman in the group. She admits she decided to make one more jump to show off in front of her curious male counterparts.

"It was really windy," she remembers. "I hooked a low turn, got a bad oscillation and hit the ground so hard. I broke and almost shattered my right ankle. That night I woke up in a hospital. The doctor said it was severe. I couldn't believe it; things had gone from okay to terrible. The next day they operated. I woke up in a drug haze, and my first visitor was John Odegard. Out of his own pocket, he'd gone and bought all my books. He said, 'Here's some reading material. Get a head start on your classes. Just pay me back when you can.'"

Weeks later, when she could walk again, she went out to the airport on the day students were assigned to instructors. But she wasn't assigned a single student. She asked the chief flight instructor, Lee Barnum, why. He told her abruptly that they'd been assigned to other instructors.

"I had this cold feeling I'd been frozen out," she recalls. "I went back to campus and John said, 'How's it going?' I told him I didn't have any students. He said, 'What?' He stood up. He was very angry, instantly mad. He said, 'You stay right there, I'll handle it.' He dropped what he was doing, ran out to his car and took off for the airport."

At the airport, Odegard learned from Barnum that the other instructors had

The modest airport facilities for Odegard's program, starting with just a corner in an old barn, were gradually upgraded through various money raising schemes. Here, work is underway in the early 70s' on expanding the ramp at the UND facility at Grand Forks Airport.

The original barn (upper right) served as the Odegard School's first hangar, with a ramp eventually built out from it to accommodate the growing fleet of aircraft. Most of the planes here in this shot from the late 70's are Cessnas with whose executives Odegard established a close relationship.

made their opinions known: none wanted Haley to work there. Barnum had said something to the effect that while he couldn't fire her, he just wouldn't give her any students. Odegard proceeded to read the riot act at high volume. Recalls Haley: "He said he'd hired me in good faith and would hear nothing of this. He didn't want them to give me a hard time. He used the term 'blatant unfairness.' Then he told the chief pilot, 'I'm ordering you to give her a shot.'"

The next day she had her students.

By then, Odegard had developed a variety of subtle approaches other than flying to ready his students for the real world. Often he would send one of them to Tom Clifford's office in Twomley Hall to request some sort of favor. Jim Bunke remembers being sent on several such missions.

"John would say, 'I really think you should make an appointment with the president's office, and go over there and state the case of the flying team and how they represent the university and that we need some support.'"

Once, on his way to Clifford's office, Bunke had an insight. "I decided that Odegard had almost certainly called Clifford to tell him, 'Bunke is on his way over to ask for this, this and this.' Let me know how he did.'"

Yet, even deducing that, Bunke still appreciated the lesson.

"I always thought I was fortunate that I related to John so well," says Bunke. "I felt like I sort of got it, got the inspiration. It was available to everybody, but if you got it, well, you were able to recognize that this was good stuff."

Like the time Bunke was assigned to drive several visiting dignitaries from campus out to the Grand Forks Air Force Base for an air show. Odegard had dreamed up an international atmospheric conference and, in grandiose fashion, had gotten the North Atlantic Treaty Organization (NATO) to be one of its sponsors. The dignitaries in Bunke's van included Harlan Cleveland, the erudite author, future university president and, at the time, American ambassador to NATO. Under a time constraint, Bunke and his passengers headed for the airport but soon ran into bumper-to-bumper traffic, crawling slowly along Highway 2. It was the only road to the base, and everyone in town was on it, heading for the air show.

"I'm thinking, 'I've got Harlan Cleveland in the van and John's waiting for us,'" Bunke recalls. "I was thinking, 'I'm never going to make it in time. What would Odegard have me do?' So I just pulled off on the shoulder, turned on my flashers and drove all the way to the airport. You talk about proud? When I told John what I'd done he said, 'God that's good Bunkerdoodle! That's exactly what you should have done.'"

Little victories like that built his confidence, says Bunke. He remembers the time he and several students were at an NIFA competition in 1978 at which American Airlines was going to present an award for the best safety presentation by an aviation program. At the last minute, Odegard grabbed Bunke and told him to

Photo by Ken Newton

Jubilation

A jubilant Jim Bunke, captain of the University of North Dakota flying team, accepts the 20th annual American Airlines safety award from J. C. (Jack) Callaway, a pilot simulator instructor at the Flight Academy. The occasion was an awards presentation at the 30th annual Tournament of Champions and Air Safety Conference sponsored by the National Intercollegiate Flying Assn. (NIFA). AA annually gives the award to the school that shows the most outstanding safety record and demonstrates a capable approach toward maintaining it. The University of North Dakota had compiled an impressive 19,000-hour, accident-free record. At left is Captain D. E. (Bud) Ehmann, AA's vice president, flight.

The wide grin on Jim Bunke's face as he accepts this trophy is directed at John Odegard who, only hours earlier, challenged him to write a summary of the safety program at UND. Bunke's impromptu report won first place in the American Airlines sponsored competition.

make the UND presentation later that afternoon. Bunke panicked. "I told him I had no information. And he says, 'But you know where to get it.'"

Bunke wasn't so sure of that. He scrambled all day to amass data about safety in general and in particular at UND. The story of what happened next is best told by viewing a photograph of Bunke taken later that day as he accepts the award from two American Airline's executives. The wide grin of amazement on Bunke's face, mugging for Odegard off camera, seems to say, 'Can you believe this? You just assigned it to me four hours ago, and we won!'"

Odegard was pleased, but maddeningly, not surprised.

"You get the assignment and, with it, the joy out of his complete confidence in you," says Bunke. "But you also get this pressure: He wants to *win it, too.* That was typical Odegard."

Take calculated risks.
That is quite different from being rash.
George S. Patton

Chapter 10
Everyone Lean Back

J ust about everyone who knew John Odegard seems to agree on two things.
One: He succeeded in developing the bustling aerospace school that bears his
name because he took risks. Two: He was a superb pilot.

The first is covered by Kent Lovelace who expounds further on the day he tried
to step down as coach of the flying team but instead came out whistling.

"From an historic standpoint, John in many ways was my Robert E. Lee," says
Lovelace, a diehard Civil War buff. "He was bold; he was a lot of initiative. He'd
look for opportunities. He took risks. I told him in 1989: 'John I can't keep
coaching the flying team. My second child is on the way. It's too hard on my wife,
the time away from home.' I said, 'I'm pushing it. It's getting risky.' I remember
him sitting at that desk in his office and looking up at me and saying, 'Kent I've
been pushing it for 20 years.' Like Robert E. Lee, he was taking risks, because there
was the potential for opportunities in those risks."

A late-arriving observer might ask, well, if he was a risk taker on the ground,
was he also a risk taker as a pilot? For it seems that almost everyone who flew with
him has a hair-raising John Odegard flying story. And in the telling, the word 'risk
taker' either comes tumbling out, or to mind: White-knuckle tales of flying through
storms, sometimes upside down; stories of near-fatal equipment problems resolved
deftly in mid air; memories of pushing the physics envelope that says this amount
of gas at this amount of speed gets you this amount of miles and no more.

Of course, it's likely that any pilot can trot out a handful of exciting escapes
and near-misses and nick-of-times. No one wants to hear stories that say "We took
off, we flew, we landed." Yet in reality, most flights are uneventful. That's what
flying is supposed to be about.

Some of the stories in the Odegard catalogue seem a bit shaggy, the kind that
are retold often enough that details begin to drop quite neatly into place as if to fit

a theme. Yet hearing these Odegard yarns, one tends to think that, at the very least, it doesn't sound as if he was a conservative pilot. Many friends and colleagues respond to this statement with a defensiveness that essentially says unless you flew with him on that day, in that situation, and unless you are a pilot, a good pilot, you wouldn't understand. In landlubber talk: You really had to be there.

Fair enough. But before the selected reel of Odegard flying stories unwinds, one might consider this thoughtful reply from former NASA astronaut Jim Buchli who flew with Odegard quite a bit.

"You can tell if someone is a good pilot in the way they conduct business in the cockpit," he says. "How they go through a checklist, what they look for and anticipate to stay ahead of the scenario. You can tell someone who is ahead of the game. You feel comfortable in terms of their decision making process. How they methodically go through the different phases of flight: take-off, landing, how they talk on the radio, how they talk to each other. Do they make good decisions? Are they very relaxed in the cockpit?

"I think John was very comfortable in knowing what he could and couldn't do in an airplane. I wouldn't consider him a conservative pilot. He knew the envelope of the airplane and was not afraid to use it. But he wasn't unsafe either. I just think he was a very good aviator.

"You can deal with aviators unsure of themselves, those who are a little nervous and who second guess themselves or who have problems making decisions. The same goes with experience. Some of it is just them. You can fly with guys who are very confident in what they do, yet they haven't got a clue how close they are to the edge of the envelope. Those are guys who can hurt you; they don't know how close they are to a limit. Or you can fly with some who are comfortable but very conservative. They never fly the plane beyond 2 to 3 hours left in their fuel reserve. They say, 'Why should I do otherwise? I'll stop early rather than late, fill up and get there with plenty of fuel.'

"Then you've got guys who say, 'I know what my minimum fuel reserve is, and I'm not going to violate it, but it's going to be further down the road before I refuel.' All of that is fine, but the key is situational awareness. John was very good at SA. You might think he stretched fuel limits more than someone else, but I've never been uncomfortable with him. I considered John extremely competent. He knew what he was doing and was comfortable flying within the envelope but not outside the envelope. Was he a risk taker? I don't think so. I think he was very experienced. Did it translate to what he did on ground? To some degree. He had an entrepreneurial bent. He was not afraid to look at something that was new. To think out of the box. More often than not you find folks who stop short because they can figure 20 ways it won't work. John was all about 'Let's figure out what will make it work.'"

That being the long of it, Don Johnston, the 747 captain at Northwest, offers the short of it: "There's a lot of measurements on how to tell a good pilot. One is that you're a good pilot if you survive."

Quick survival story number one:

Hal Gershman, the Grand Forks city council member, remembers a flight with Odegard to Bismarck in the venerable old single-engine Mooney Executive. Having just taken off and reached 800 feet, the engine stalled.

"I'm thinking," says Gershman, "'There's a farm down there, and I'm going to buy it.' But John never panicked. He was cool as a cucumber. He got the engine started again and, still sputtering, he brought the thing in. To him it was like your windshield wipers didn't work for a second."

And another from insurance executive Dave Vaaler:

"I was scared a few times flying with John but not because of anything he did wrong. Well, for example, one time we went hunting in Chamberlain, South Dakota. There were four of us in the plane. I was in the right seat. It was cloudy with a low ceiling. He says to me, 'Now you've got a job here. I want you to do this…if we have to abort this thing, you're going to have to do this.' I was a little nervous that I would know what to do and do what I was supposed to do. We're going down there through the clouds and all of a sudden we broke out and there was the runway. He put it down beautifully."

Jim Odegard says his brother was a good pilot because he learned from mistakes. Take the time, he says, in John's early years at the university when he was still doing crop spraying. The sprayer's job takes place early in the morning and late in the evening when the wind velocity is usually lower and won't blow the spray askew. One morning, John got up at daybreak and was very tired.

"He made a few passes over the field," says Jim. "The sun was just coming up; it was warm in the cockpit. He dozed off. Next thing he knows he wakes up and here are the trees at the end of the field. He pulled back and as he's clearing the trees, some branches hit the spray boom. He told me he went up around, made a pass and landed and said that's enough for today. He used good common sense."

Gerry Skogley, UND's former vice president of finance, often sat in the right seat with Odegard on university flights. "I would have flown anywhere in the world with him," he says. "In the context of taking risks—no, he always knew what he was doing. He was always in control. We flew through a thunderstorm going to Bismarck one day. A young air traffic controller took us around south and put us in the middle of a storm. There was hail bouncing off so hard it took the paint off the plane. It felt like we were upside down two or three times. It was not a good thing. It was the only time I saw John's knuckles go white. I was frightened. I never was concerned about crashing. I just didn't like hanging in my seatbelt. But John knew what to do. He was calm, but I don't think he'd ever been there before. He was very skillful and never missed the outer marker."

One of Odegard's most important passengers, Tom Clifford, says he felt so at ease with Odegard as a pilot that he often fell asleep, even in the middle of storms. Once, he recalls, during the days in the late sixties when he and Odegard were crisscrossing the state and visiting members of the Board of Higher Education, they were flying home to Grand Forks from Bismarck.

"It was snowing so hard you couldn't see your hand in front of your face," says Clifford. "John said, 'Look out there and see where we are.' I did and I found this little church down there and an open road leading into the airport. We came in and landed, but still couldn't see much. John radioed the tower and said, 'Where do you want us to go?' And they said, 'Don't land.' He said, 'I am landed; I'm parked right outside your window.'"

Clifford remembers other trips during those hops from small town to small town in the old Mooney. Sudden bad weather often forced them to land in odd places.

"Sometimes we landed in fields. I never had any concern. I'd go to sleep. I'd wake up and look around and see what was going on. I can remember flying over Bottineau; we got down low so the prop would blow the water out of the ruts in the dirt road, so we could see how deep they were. Then it was like a carrier landing. Just 'Whoom!' Once we had to get a farmer to pull us out with his tractor. John was a marvelous pilot, because he loved it. And by the way, we never lost a vote on the board. I think they liked us both. There was a certain boldness that attracted them."

Exciting stories all, but what about risk taking?

Don Johnston recalls a raffle the Flying Club held in the late sixties. Students had set up a ticket booth in the parking lot of a shopping center on Columbia Road. To drum up ticket sales, Odegard had arranged for a club member to taxi a Cessna 120 to the lot from a field on South Washington Street about two miles distant. After a brisk day of ticket selling, Odegard volunteered to taxi the plane back out to the South Washington field.

"But why taxi it," asks Johnston with a smile, "when you can just fly. The wind was right. So he took off. He got about 100 feet in the air before the engine quit. Some fuel selector switch had been flipped off. He immediately flipped it back on. The engine recovered, and everything was great. That was one of those things, kind of an impulsive decision. Sometimes impulsiveness is not the best trait for pilot. There's other stories like that. But he never panicked in those situations."

In fact, three of the highest ranking ex-military men Odegard hired over the years, say he would have made a fine fighter pilot.

"John was very disciplined in flying," says retired Air Force Maj. Gen. Darrol Schroeder. "That's why he stayed a very safe, good pilot. A few of us with military background thought John would have made one hell of a fighter pilot. In the old sense of the word, a fighter pilot had a different method of approaching things than a bomber pilot did. The cream of the crop were fighter pilots. They didn't pick a reckless person, but someone willing to go to the edge. John had that type of makeup."

Retired Colonel George Hammond, the Vietnam fighter pilot, agrees. "Some probably thought he was a risk taker. Give you an example. We were in the Citation jet going to Seattle. We came past Mt. Rainier and were going into McCord Air Force base. John was making the approach. As we came across the mountains and were letting down he said, 'Let's put it on idle and see if it can make the runway.' We were only 15 miles from the airport. He put it down on the runway right past the numbers and never had to add power or anything. He was an outstanding aviator."

Just how difficult was that maneuver? Al Palmer, a recently minted Brigadier General in the North Dakota Air National Guard recalls a day with Odegard in the Citation when he performed the same experiment—this time on the way to Bismarck.

"He pulled the power back to idle at 22,000 feet, 60 miles out," says Palmer, "and he landed in Bismarck. I'm thinking that's pretty damn good. It's managing energy. The closer you are to the ground, the easier it is to make happen. When you do it from 22,000 feet 60 miles away and with no power, you're managing that energy all the way down. If you're too slow, you land too short; with too much energy you can't land because you're too high. I always remembered that. John was a natural pilot.

"I remember we were going into Spencer, Iowa one day in the Citation. It was early one morning at 6 a.m. They'd had an ice storm, and the runway had ice. I was flying the jet. I brought it in and touched down, and I couldn't get it to stop. I was uncomfortable, so I just fired it up and went around. I told John I didn't feel comfortable landing. So he took it, and he brought it in and landed that thing. He could do that."

But for a really hair-raising Odegard yarn, Don Smith may take the prize. He still isn't too happy about it.

"When John flew anything scheduled, he was very disciplined," says Smith. "But he liked to be on the edge. One time John, George Hammond and I picked up two Sundowners in Wichita, Kansas, from Beechcraft. A fellow in Minot was going to buy the planes long distance through a dealer in Bismarck, and then lease them to UND. We get down to Wichita, but this guy hadn't gotten the money to the Bismarck dealer in time. We couldn't take the planes until they were paid for. John was furious. He got the dealer on the phone and talked him into wiring the money, so we could take the planes. The money came in around five o'clock in the afternoon.

"The deal was George Hammond would fly the first plane. He spent the whole day reading the manual. Meanwhile John was upset, running around trying to get this settled. He had a meeting the next day in Grand Forks, and I did too. Despite the fact that the weather made it all instrument flying all the way back, we got the planes and took off. And I mean it was bad. Pouring rain even at high altitude.

"George was about five miles ahead of us. About 80 miles out from Grand Forks, John tries the landing gear, and the gear won't go down and lock. He tried again, and it didn't go down. He hadn't read the manual on how to drop the gear by hand. He's looking around for the manual, and I say, 'John why don't we try one more time?' We try again, and the gear goes down, but it doesn't lock. So he got his flashlight and said, 'You fly Don.' So I'm flying in this terrible storm on instruments. [Note: Smith never had a pilot's license] He's got a flashlight and is reading the manual. Finally he works it, and there's a groan, and the gear locks. At that point I would have sacrificed a little speed and left the gear down. But John retracted the gear. We got to Grand Forks and by God, the gear wouldn't lock again. We went through the same thing, and it finally locked. We landed. That's trying the edge."

Odegard was aware of his reputation as an edge-rider; in many respects he didn't care, because while *they* thought he was taking a risk, *he* knew he wasn't. During an interview in 1979 he told Dick Youngblood of the *Minneapolis Tribune* the Tom Clifford story—landing in the storm just before the control tower told him not to land. He followed it up with an immediate appeal to Youngblood: "Don't put that in the paper. I'd flunk any student who pulled a stupid stunt like that."

In terms of the line you don't cross as a pilot, Odegard tended to draw it between himself and his students. While he was not exactly saying to them, 'Do as I say, not as I do,' he knew that his wide-open style of flying wasn't the kind of thing a beginning, even a practiced student, should attempt—kids, don't try this at home.

People will say about Odegard that when he walked into a room he instantly became the dominating presence. In a space as confined as a cockpit, that same mix of reputation and charisma could easily intimidate a student. One former student who knows that first hand is Diane Odegard, who took flying lessons from John and even accumulated some hours.

"He was the teacher, and I did not do well," she says. "You should never have your husband teach you to fly. We had a little tiff about something after we landed the plane one day, and I didn't take any more lessons."

Flying wasn't really important to her anyway, she says. Not so for Jerry Murray. One day, he took his special closeness with his neighbor a little too lightly. He'd been flying with Odegard and was tasked to land their small plane at Grand Forks Airport.

"I remember being on final approach," says Murray. "We'd been cleared to land, and a Northwest 727 was landing just ahead of us. I made some comment to John about the 727, and he just barked back at me, 'Concentrate on what you're doing!' It was the first time he'd ever barked at me. But I never forgot that in all the years I flew professionally at Cessna. As I started flying bigger and bigger planes I always remembered John barking at me. When I started doing some instructing

When Odegard's Center for Aerospace Sciences took delivery of a new Beechcraft jet in the early 90's, Beech pilot Ron Gunarson (right), UND Class of '88, brought the plane to Grand Forks where Odegard himself took it for an immediate test flight.

An avid hunter, Jerry Murray (left) had plans to become a wildlife biologist until John Odegard moved in next door. Hooked on flying, Murray graduated from Odegard's program. While working in Texas at SimuFlite, he and Odegard found time for some hunting.

with young people, I passed it on. Not the barking. But once you're cleared to land, there is no chit chat."

On another occasion, Odegard took Murray up in a Cessna 150 fitted out with a special airframe so it could withstand aerobatic maneuvers. He performed every twist and loop in the book—literally. They were all tricks he had taught himself as a young man by taking a plane up, strapping a book of maneuvers to his thigh, and following the diagrams as he went through each one.

Several lessons took hold on that flight, says Murray. "I never got sick. Any good pilot can make a passenger sick if they want to. It's easy to do. John wasn't like that. With a good acrobatic pilot, you can know you're upside down but not feel like you're upside down, because you keep the forces constant. John had a little glass of water. He put it on the dash and said, 'I won't spill a drop.' We never spilled a drop. The forces are always constant, so you're always on that same G level."

Toward the end of the lesson, Murray felt he had stumbled on a profound truth.

"It was obvious that flying was the best part of his day," he says. "He was a completely different person in the cockpit than he was in class or the office. When you fly, you block everything out and concentrate on what you're doing. You have to. John was at his most serious when he was flying. Complete concentration on what he was doing. He was very proficient at it. He was everything I thought he would be as a pilot."

In the air, Odegard never stopped being an instructor. Many of his students who'd already earned their pilot's license and advanced certifications, tell stories of piloting a plane with Odegard. He would take pieces of paper and cover up various dials on the instrument panel. He would then ask a series of rapid fired questions about what actions would be taken if this or that happened.

"He could be disconcerting to fly with," says Bob Muhs, the former student and now Northwest executive. "It was always a check ride. You didn't want to screw up. The minute you got in the plane you just had to be ready, because he wanted you to be ready."

Classic was the time Odegard, Muhs and two other students took a V-tail Bonanza to Denver for a conference. Odegard placed Muhs in the left seat. From the moment they were all aboard, he began asking Muhs questions about the fight plan. "It was just non-stop talking," says Muhs.

As they approached Denver, Odegard asked Muhs a question about the weight balance in the V-tail—a plane none of the students had flown before: Is the center of gravity going to be aft or forward?

"And I'm thinking," says Muhs. "I said we're going to burn aft. Which was right. It was a lucky guess. He said, 'We're real heavy, so when you land with an aft center of gravity the plane is going to porpoise to the right. You want to come in a little bit and don't flare as much as you normally would with less weight." But as Muhs

Bob Muhs, shown here at his 1997 wedding, became an Odegard favorite during his undergrad days at the University of North Dakota, even babysitting for John and Diane. Helped by Odegard's contacts with Northwest Airlines, Muhs landed a job there after graduation and remains an NWA executive today.

The astronaut Jim Buchli was raised in Fargo and had already flown a shuttle flight before he and John Odegard met. As with everyone, Odegard immediately impressed Buchli who became a close friend and consultant over the years and a favorite fishing companion.

brought the plane in, he flared normally, "which was much too much, and we did the porpoise to the right and got onto the ground."

Odegard said, "See? You gotta carry a little more speed and make just a slight correction."

Three days later when they left Denver, Odegard picked up the instructor's role again, putting Muhs back into the pilot's seat. As they taxied onto the runway they fell in behind a 737. "John said, 'See how when he taxies he does a little sideways kilter there on the taxiway? That means you're too close." Muhs nodded and dropped back.

Moments later, as they prepared to take off from the Mile High City, Odegard took Muhs through a density altitude drill, emphasizing the airplane's weight and their 5,280 feet above sea level. "You're going to take a lot of runway," he said.

"So we take off," says Muhs, "and we get off the ground, but just. We were getting no climb at all. He's over there in the right seat sucking it up. Finally, we get a positive rate, the gear is up, and we're just slowly climbing. It takes us forever to get to altitude. We were trying to make it back to Grand Forks non-stop. He says, 'Do we have enough fuel to make it non-stop? Where would we stop if we needed to get fuel?'"

Muhs thought about it. Odegard said, "Well? Can we make it nonstop or not?" Muhs nodded, "Yeah, we can." Odegard smiled and Muhs realized once more that he'd given the right answer.

But the lesson wasn't yet over.

"We were cruising along," says Muhs, "and Odegard brought up the center of gravity issue again. 'I want to show you something,' he said, turning to include those in back. 'Everyone sit still. Bobby, take your hands off the wheel, foot off the rudder.'"

Muhs did so, noting that the plane didn't have an auto pilot.

"When I tell you to lean forward," said Odegard, "lean forward. Okay, lean forward."

Everybody leaned forward and the plane pitched over in a steep nose down attitude.

"All right," said Odegard, "everyone lean back."

They did it, and the plane righted itself. They repeated the exercise several times. Muhs was surprised that they never lost much altitude.

"So now," said Odegard, "you got the feel of what this plane will do with heavy weight and the airplane burning aft. Remember what happened in Denver. When we land in Grand Forks you now know how sensitive this plane is. You've got to have a small input. If you carry a lot of speed and make those small corrections you don't have to worry about the oscillation. How much speed do you think we need?"

He reminded everyone that the normal approach of 90 knots wouldn't be enough given the weight-to-CG issue. "You're going to want to carry about 30 extra knots of speed."

As they came in, the wind was stiff.

"He's sitting there," remembers Muhs, "the classic instructor. So I land. I'm coming in about 125 miles an hour, making these small corrections. We land, and of course we're going fast for a Beech. The runway is going by fast, and we get the brakes on. He says, 'Yeah! That's what you need to do!' And we were done for the day, and I was soaking wet."

But a more confident soaking wet. As Jerry Murray puts it, "You do gain a lot of confidence in everything you do in life after you've learned how to fly. You've mastered something very few others do. You learn also that if you're going to be any good at it you've got to be serious about it. John told me, when I started flying jets, you've got to start thinking 500 miles an hour. You have to think ahead of the airplane, about how fast you're moving across the ground. You have to accelerate your thought process. Your judgment and decision making has to be in synch with the plane."

Much harder, though, was bringing that 500 mile-an-hour thinking into synch with the plane of the earth. Odegard's expectations for perfectionism on the ground in North Dakota were thought by some to be unreasonable. Yet in his mind he had already achieved it in the air, where perfectionism wasn't simply possible, it was as important as breathing.

"Somebody gets into trouble,
gets out of it again.
People love that story.
They never get tired of it."
Kurt Vonnegut, Jr.

Chapter 11
The Woodshed

In an ironic twist on the what-goes-around-comes-around game, John Odegard, in 1974, hired the man who had once revoked his license for shenanigans in the air. Les Severance, the former chief of the FAA's General Aviation District Office in Fargo, already had been so supportive in helping Odegard get the proper government approvals for his fledgling program that Tom Clifford awarded him the UND President's Medal. Severance wore several hats for Odegard, acting primarily as the ongoing liaison with the FAA. But he also supervised the maintenance operation at the airport under Col. George Hammond. He helped to acquire new aircraft, gave students their ultimate check ride before earning their license and coached the UND Flying Team.

With the official establishment of a department of aviation in 1968, the need for a flying *club* faded. By the time it was supplanted by an official UND Flying Team, Odegard's old flying club had helped 300 of its members get their pilot's license over the years. According to a story in the *Grand Forks Herald*, the club had grown since Odegard's arrival from 14 members to 104. The story also noted "there are seven girl members."

One other item of note in that article: The official UND Flying Team won the first place prize in 1971 in the Region 2 air meet of the National Intercollegiate Flying Association. Here was an early example of Odegard's national recruiting strategy paying off. He and Gary Kiteley had arranged for the University Aviation Association to break into regions. Instead of all the glory and prestige going to just the one school that captured the national NIFA title each year, a handful of schools now could claim a regional title and boost its image. By 1973, UND's Flying Team took first place in the NIFA regional, qualifying the university for the first time ever to compete in the national championship. Even though it didn't

Odegard was lucky to get both of these men into his program. Les
Severance, (left) a former FAA executive, had once grounded a teenaged
Odegard for stunt flying. He later served as liaison between UND and
the FAA. Al Palmer, right, rose from Odegard flight instructor to the key
airport post of director of flight operations.

In spite of all of his jetting around the globe searching for business
opportunities on behalf of his school, John Odegard remained very much
a family man, keeping in constant communication with Diane and his
two children, John Jr., and Stephanie.

win, the map of the college aviation world now officially included a new power.

It didn't look like much of a power, though. The Department of aviation in 1972 had the Mooney, 3 Cessna 150s, a 172A and a 310 that had been donated by the Air Force as surplus. When Jerry Nelson arrived on campus as dispatcher in the fall of that year, he was shown to his office in a small house trailer that had been installed in front of the old barn of a city hangar.

"My office was the kitchen," he recalls. "The instructors had bedrooms."

He soon learned that the essential job of the dispatcher was to work the line—pull planes out of the hangar, start them and *then* dispatch them. He even changed oil in them.

Recalls Nelson: "I'd go home at night and say, 'Why am I doing this?' But next day you were back out there. Everybody worked hard out there."

And getting planes out for students on cold winter mornings could be rough work indeed. The planes were crowded into a tiny corner of the unheated hangar, cheek by jowl up against the rudimentary maintenance shop. First task: fire up the space heater, warm up the cockpit of a plane, pull it out onto the ramp and take one of the blankets Odegard had filched from somewhere and throw it over the engine cowling to keep it warm until a student arrived. Then start the process again with the next plane.

While the school still operated its DC-3, Nelson and others would similarly pull it out of the hangar and start it up. But because the warped hangar doors wouldn't close, the backwash from the props blew the winter wind right through the tarp that had been hung around the UND quarter of the hangar.

"There was some heat, but every time the wind would blow, the air would come in," says Frank Argenziano. "I mean, that was miserable."

Argenziano, a native of the Fargo area and an NDSU grad, had been an airplane mechanic in the Army. Les Severance hired him to run the maintenance shop and help him get it certified by the FAA.

"Les was a big, steady man," says Bob Reis, Odegard's first director of fiscal affairs. "He was one who put a lot of thought into whatever he had to say. As a result, we didn't spare dollars on maintenance. John did not want any accident to be caused by mechanical failure. Sometimes we'd struggle with costs on maintenance, but John would always say he wanted never to be afraid of getting into an airplane maintained by us. Never. So we did not skimp on things like running a tire an extra 20 hours. It was replaced."

At the time, Odegard had about 50 students—that is, students majoring in aviation. But there were several hundred other students from other majors, or veterans with G.I. benefits, who simply wanted to learn to fly. They would sign up for ground school classes and take the lab which amounted to flying lessons taught by Lee Barnum or one of his instructors.

The overload on the small, inadequate facilities was felt by all.

"When I first got there," recalls Jerry Nelson, "I said to myself I don't know how we can go any further. We had all these old planes. But John kept pecking away. We'd go to lunch, and he'd tell me the things he was going to do next. By golly he did them and got them done. I've never seen anything grow as fast as that did."

The first expansion was an inevitable no-brainer: The remaining three-quarters of the city hangar were soon encroached upon by flight operations, pushing the city to officially grant them the entire structure.

Next, Odegard plied his charisma on the owner of the Candor construction company in Fargo. They built a modest, concrete building next to the city hangar, which nicely housed the maintenance shop. The building was then leased back to the university. Odegard would eventually buy both the Candor building and the city hangar outright.

From day one in 1968, Odegard fully expected to run both ends of the aviation equation: the academic program on campus—including ground school, aviation business courses and the core liberal arts classes everyone took—and the airport operation. But for several years into the early seventies, Odegard's campus critics mounted a campaign aimed at taking the airport operations out of his control. Unwittingly—or more likely, wittingly, but without giving a damn—Odegard provided the fuel his critics needed.

The first thing he did to demonstrate disregard for rules everyone else had to follow was the little matter of that old barn of a hangar at the city airport. To get his original one-quarter of the drafty, leaking structure, Odegard entered negotiations with the Grand Forks City Council and the director of the airport and finally signed a lease. Two minor details: He had no authority to commit the university to a lease. And even if he had, there wasn't a cent in the budget to pay for it.

"We found out after it happened," says Gerry Skogley, at the time the comptroller of the university. In 1970, when Clifford became president, Skogley succeeded him as vice-president for finance. "John didn't have any idea where the money would come from to pay the lease," he says. "The president had some discretionary monies and typically that's where it came from. We'd take it away from somebody else or some other purpose and bail John out.

"I would say to him, 'Why did you do that?' He would say, 'Well, I thought it was the right thing to do and it needed to get done.' And he'd say, 'I'm sorry.' And Tom would pat him on the head a little bit and say, 'We'll get it worked out John.' And we always did. But it was terribly frustrating."

Skogley wasn't alone in the frustrations-felt department.

"John was highly thought of in the aviation industry," acknowledges Lyle Beiswenger, Skogley's successor in the eighties. "But he had an inability to handle finances. He was too much off the cuff, doing things and worrying about paying for them later. One of John's philosophies was that it was easier to ask forgiveness than permission. If he overspent on an account, well, Tom or Lyle or somebody will pick up the difference."

As noted earlier, one of Tom Clifford's most oft-practiced philosophies held that it was easier to ask forgiveness than permission. His protégé learned that lesson quickly. Clifford's successful career was rife with steps taken that were against either the letter or spirit of some rule or law—from bold tactics in the Marine Corps, to his calculated end runs as a college administrator and president. The three most prestigious and grant-endowed elements of the university today—the School of Medicine, The Odegard School and the Environmental Energy Research Center—most likely would not exist without rules having been bent and sometimes broken by Clifford. While Odegard clearly learned the strategy from the master, one key difference between him and his mentor was that when it came time to actually ask forgiveness, Clifford's engaging sense of humor and Irish charm made it very difficult for a state legislator or board of education member to remain angry. Odegard, always impatient to get things done, was usually less eager and skillful at playing the diplomacy game.

From his point of view, the flouting of rules was precisely what was needed. "The accounting crowd, they're the worst thing that can happen to an aviation program," says decorated Vietnam fighter pilot George Hammond. "They don't understand, and they can't count above 100 dollars. They don't trust fliers. They think we're all wastrels. They thought John was a little wild and reckless with money, but he made it happen. I'm sure today they're happy to have the school."

Even so, Odegard's contrition rang hollow on some ears, especially as it was almost always followed up by repeat offenses. A classic example, recited with humor by Don Smith, is the saga of the Checkered Cab.

"We had no transportation out to the airport," he recalls. "Some kids who had a car could get out there, but others who didn't had to bum a ride. It was like a missing link. Anyway, the manager of the Holiday Inn was a friend of Dean Jelsing, our manager at the airport at that time. The Holiday Inn had an old Checker cab to pick people up at the airport. They gave that old car to Jelsing. He and the staff out there took that old Checker cab, went through the drive train and engine, stripped the paint and put on a coat of white enamel. They put slip covers on the seats and made it a real gem. On the side—this was John's idea—they put the seal of the state of North Dakota in gold leaf. Above that in dark green letters 'Department of Aviation.'

"It had no license plates. Maybe we were naïve thinking a state vehicle didn't need plates. But it wasn't a state vehicle. We'd ask a student, 'Can you drive?' If he could, we said, 'Okay, take the cab out to the airport.' If there was a kid coming back from the airport, we'd get him to drive the cab back. So this cab is now being driven back and forth. We also used it to go to the store, to pick up parts, or whatever.

"Now the manager of the Holiday Inn sees this car go by, and he wants to get some publicity out of it. So he gets hold of Dean Jelsing, and they get a picture that went into the *Grand Forks Herald*, with the manager presenting the keys to Dean."

That morning, says Smith, Loren Swanson, the university's vice president of operations and the man to whom Odegard's airport operation reported, stormed into Gamble Hall with a copy of *The Herald* crumpled in his fist. "Swannie was so upset," says Smith, "I swear he was two feet off the deck. He says, 'What in the hell are you guys doing?'"

Well, says Smith, everybody thought the state was self-insured. Swannie got even more upset when he found out just anybody could and did drive the cab.

Whether the result of this little episode could be classified as a Tom Sawyer or Br'er Rabbit strategy, Odegard got what he wanted. Swanson took away the cab and gave the program an old Chevy. "And that," says Smith, "is how we got our car."

"Would we ever have been able to get what we needed by going through the normal process?" asks Leon Osborne, one of the pioneers in the early department of atmospheric sciences. "I think the answer would be 'no.' John had to fight to create the school. His attitude: 'We're going to get this done and the hell with the rest of the university.' But we had to live with the repercussions."

One of which was the continuing pressure on Tom Clifford by his faculty to run Odegard out of town on a rail. Sometimes even the master himself was tempted.

"He drove Clifford nuts," says Bob Muhs, one of those early Odegard students. "John didn't manage money well. He just spent a lot of money and wasn't sure where he was going to get it."

Smith says such laxity often landed Odegard in hot water with the president. "Tom would call us up to his office," he recalls. "He had two vocabularies. He could either sound like a typical college president, or he could make a dock worker blush. But he always supported us. It was always push away but don't stop. He covered our butts."

Former North Dakota governor Allen Olson knew both Odegard and Clifford well. "I think there was a kindred spirit, there," he says. "Both understood business. Both were competitive, smart and savvy. I suppose Tom might have seen a little bit of himself in John, but he also saw a person who had a vision. Part of Tom's mastery was that he delegated. He put good people in place and let 'em go. Both probably recognized in each other a willingness to push the envelope. I suspect John probably pushed Clifford too far at times. I suspect there was probably a very discrete, private taking-to-the woodshed at times."

Bob Reis remembers the day Odegard hired him as his director of fiscal affairs.

"He said, 'Don't try to get in my way. Just keep me out of trouble.' John just moved fast, sometimes too fast, and he wanted me to look underneath the woodpile. The whole staff was trained to look under woodpiles because there might be something we don't want there. If it was insignificant, John would say, 'We'll take the risk.' Sometimes I'd say, 'Hey this is a problem area for us, we've got to stay away.' Sometimes he would listen.'"

When he didn't, though, even Gerry Skogley could appreciate why.

"John didn't have anything," he says. "He had no state support. He had a few dollars from students for flying lessons and that was it. The income from flying lessons was adequate to take care of instructors and maintenance. But there was nothing there to expand. He would come up with ingenious ways to figure how to get more space, more airplanes. But he was not always up-front. You'd find out about it after it happened. So you'd be picking up the pieces and putting out fires. He didn't deny doing the things. He had this dream of developing a program and he just bulled ahead and created the program and let some of the rest of us pick up the pieces along the way. He just didn't have any idea of what the big picture was in terms of the cost and how it had to be paid for and who would pay for it. He just didn't comprehend that."

Bob Muhs has a theory about Odegard and his difficulties with finances. "I will never believe," he says, "that he didn't know what he was doing on the economic side. There was no doubt he was lucky, but he created his luck. He's got an aviation school in North Dakota and there's no reason on God's green earth it should be there."

Meanwhile, Don Smith says that on occasion—usually on a Sunday when both were in the office—Odegard would pull out a bottle of Scotch and pour them each a drink. He'd then chuckle as he offered a toast to his latest successful round of forgiveness-seeking from the bean counters.

But a larger crisis loomed in those first years of the seventies that went beyond simple permission and forgiveness. Skogley had commissioned an outside study of the airport operations. It concluded that flight operations was being inefficiently run and wasn't making enough money to cover all of Odegard's airport costs. It recommended removing flight operations from Odegard's control and contracting with a fixed base operator to run the airport end of things.

"That would have been the kiss of death for John," says Don Smith.

He, along with Odegard and others, viewed the report as the culmination of a grand conspiracy against them, a plot orchestrated by faculty, administrators and FBOs. "The word was out," said Smith. "Put every obstacle against us."

Skogley discounts the conspiracy theory and says the idea was that the airport program stood a better chance of getting the funding it needed from the state without Odegard in charge. His old promise never to seek state funds meant that if he wanted a bigger budget, he'd have to persuade the deans of the other colleges to share some of the revenue handed down to them through the state.

"He was having difficulty on campus with the other deans not wanting to give him any of their money," says Skogley. "It was quite a struggle."

Separating flight operations also raised the possibility, adds Skogley, that with just the academic side of things under his watch, Odegard would have less cause "to stir the pot."

Skogley's airport recommendation, which came out just as the 1973 fall semester was getting underway, put Clifford in a tight spot. The deal he'd made with Odegard was sink or swim on your own—even though privately he'd tossed him plenty of lifelines. On the other hand, Gerry Skogley was his vice president and Clifford looked to him for much the same reason as Odegard looked to Bob Reis: keep me out of trouble. And like Odegard, sometimes he would listen.

"Every campus has people doing things where they don't want to go through the process," says Skogley. "I spent a disproportionate amount of time working with John and his people. For some, I was the bad guy. Someone had to do it. I was willing to, and Tom was supportive of my process."

Two events took place then that changed the landscape figuratively and literally. One of them bought Odegard time. The other bought him the magic seeds of an empire.

The first was as predictable as its outcome: Don Smith and George Hammond marched in to see the academic vice president, Alice Clark. They told her that if the airport was separated from Odegard, they were both quitting. The threat wasn't just awkward, it was potential for disaster. Aviation students already had enrolled in their fall courses, most of which were to be taught by Don Smith. Those students also had signed binding contracts regarding payment of fees. It seemed a huge mess to Alice Clark, but Clifford found in it the perfect logistical, face-saving excuse to postpone the airport separation for a year.

The second noteworthy event played itself out devoid of drama. It was another of those teetering, fragile moments that slide by unnoticed, but in retrospect stand out like a flying buttress holding up a cathedral. It involved a simple yes or no question that Odegard posed to a man named Patrick Hurley—a man he'd never met.

Here is the short version:

John Odegard: "Do you ski?"

Patrick Hurley: "Yeah, I do."

Today it can be argued that had Hurley answered, "no," the cathedral might have collapsed or ended up little more than a small prairie church house.

PART TWO
THE TUMBLING MIRTH

Life happens too fast

for you ever to think about it.

If you could just persuade people of this,

but they insist on amassing information.

Kurt Vonnegut, Jr.

Ride a painted pony,
let the spinning wheel fly.
David Clayton-Thomas
Blood, Sweat & Tears

Chapter 12
Click, Click, Click

John Odegard and Patrick Hurley whispered their fateful few words to each other in a council room in Bismarck, in December 1973. They'd each come to the state capitol from differing backgrounds to attend a meeting of the state's Weather Modification Association—overseen then by the North Dakota Aeronautics Commission.

In a state whose economy relies so heavily on agriculture, the need to control the clouds that breeze in from Montana and down from Canada was obvious. The desired control fell into two categories: wring as much rain as possible from those clouds, and tame them to the point where frequent storms of crop-killing hail are neutralized.

States that set up a weather modification apparatus usually place it in their department of agriculture. It was John Odegard's good fortune that in the late sixties North Dakota established its board under the aeronautics commission. Fortunate, because Odegard—as head of the university's department of aviation— was a *de facto* gubernatorial appointee to the commission. In December 1973, he was also the commission's chair.

The main purpose of the weather modification panel was to oversee various cloud seeding initiatives in the state. The idea of injecting silver iodide particles into those parts of clouds containing super-cooled liquid water was developed in the 1940s by the General Electric corporation. The particles essentially jump start and extend precipitation. They add to a given cloud's existing ice crystal formations. They then trigger an earlier and heavier rainfall than if they hadn't been there. The process that leads to the creation of large hail stones is also interrupted.

In the late fifties and early sixties, The National Science Foundation began funding research projects aimed at improving the methods and techniques of seeding clouds. Some of those projects demonstrated that cloud seeding by aircraft was the most effective format—as opposed to the burning of silver iodide on the ground and guiding its smoke into cloud formations.

The high plains states of South and North Dakota were among the first to produce effective cloud seeding results using airplanes. In fact, a Fargo company called Weather Modification Inc., with its own specially equipped planes, began its research into cloud seeding as early as 1961. It continues its successful operations today.

For scientists, the seeding process was natural and harmless, though some of the stolid old farmers and ranchers who would benefit, viewed it as unnatural and harmful—much as skeptics have viewed fluoridation of public drinking water over the years. John Odegard saw only a natural opportunity. He had nursed the idea for some time that, since cloud seeding required airplanes there had to be a way to turn it to his advantage. Typically, when scrutinizing an opportunity whose potential seemed vague to others, Odegard would shake his head and say, "There's got to be a pony in there somewhere for us."

Indeed, earlier in 1973 a small pony had emerged when the National Science Foundation became interested in funding research and educational programs in weather modification. It was a time when weather modification had become less experimental, more accepted and in need of thorough professional study. When North Dakota's Congressional delegation learned that federal monies were being awarded they pressured the NSF into cutting off a small slice—$235,000—for UND. As a result the university's department of geography in the College of Arts and Sciences was slated to teach classes in meteorology. Odegard's crew would train students in the techniques of flying a twin-engine, cloud seeding aircraft. It was to be the first such pilot-training program ever in weather modification flying. Already some of Odegard's students who had been certified as multi-engine pilots were doing part-time cloud seeding in the western part of the state for Fargo's Weather Modification Inc. One of them was Jean Haley, who became the first woman in the country to qualify as a weather modification pilot.

It was all very nice, but Odegard wanted a larger pony. In fact, he was counting on a herd. The rumbling of distant hooves was heard in May of 1973 with a directive to the Bureau of Reclamation by its parent organization, the U.S. Department of the Interior. The stated mission of the BuRec—as it was known in official shorthand—was water resource development in 17 western states. The Interior Department directed it to develop a "comprehensive program of cloud seeding experimentation in the High Plains Region."

The program would study not just cloud seeding, but the downwind effects of cloud seeding. It talked of using new equipment, such as digital radar towers that could follow researchers from spot to spot.

In a conference held in the spring of 1973 at the Denver office of the BuRec, educational and governmental representatives from 10 high plains states came together and agreed on many details of what the program should do. They proposed a list of seven possible sites for three federally funded field research offices—one each for the Northern, Central and Southern high plains states. One of the seven suggested cites was Williston, in western North Dakota.

In the early selection process, several key scientists at the University of Montana volunteered to go to work for the BuRec if it would set the northern states regional field office in Miles City, MT. The bureau agreed, and turned its attention to selecting the central and southern offices. Even as it did, the bureau worried about the reaction of the North Dakota Congressional contingent to the bypassing of Williston. There existed the strong possibility that by choosing Montana over North Dakota—a state whose Congressional representatives had always voted favorably for BuRec appropriations—the bureau could be seeding the clouds of its own demise.

It was clear that the bureau had to find a political solution, some way North Dakota could play a significant role in the Miles City piece of the project without it looking like, well, a political solution. To resolve the issue, the bureau turned to an unlikely bureaucrat, an Air Force veteran with degrees in meteorology and public administration. He was also a man who had become something of an embarrassment to them by robustly refusing job offers in the bureau's Washington office. Not so surprising, really, since he'd been born and reared in North Dakota.

His name was Patrick Hurley.

Hurley worked out of the Denver office of the bureau as the head of planning and special projects. By heading up the site-selection process for what had become known as the HIPLEX program he knew that the right scientific credentials were important, but only as important as the political considerations.

"The BuRec was a very political group," says Hurley. "It was one of the pork barrel agencies, to put it politely. We were very involved with congressional delegations."

Hurley was no disinterested observer in his assignment. He was from North Dakota and thought North Dakota ought to get some of the HIPLEX money. He just didn't know who it should go to or for what rational reason. The deadline for determining such things was less than two months away. Hurley flew from Denver to North Dakota to meet with Harold Vavra, the head of the Aeronautics Commission. Together they toured the state, considering various scenarios, none of which sounded plausible enough to Hurley. The tour ended with a large conference in Bismarck.

Hurley opened the meeting with a speech about HIPLEX, putting emphasis on the need for mobile radar which could move about to examine moving cloud formations. When he took his seat someone else got up to speak. Hurley's attention wavered. He started fiddling around, bored, shifting this way and that in his seat.

Out of the corner of his eye, he noticed that the fellow sitting next to him, someone he knew only as a member of the commission, looked equally bored and itchy.

"He looked at me," Hurley recalls. "I had a belt buckle that said Vail. He leans over and says, 'Do you ski?' I said, 'Yeah I do.' We got to talking. He said, 'You wanna come over to Grand Forks and see the hockey game? UND's playing Denver.' I said, 'Yeah, that would be great.' He said, 'Okay, we'll fly to Grand Forks tonight and watch the hockey game.' I said, 'Sure.'"

That night when John Odegard took off from Bismarck, his fledgling school—while not as narrow in focus as its critics made it out to be—still concentrated primarily on training students to fly and to learn the intricacies of the aviation business. It had no research component, no academics on staff. It had to share the NSF education grant with the skeptical geography department, and because of the Skogley airport study soon might be unable to control its own airplanes or flight instructors.

By the time he landed in Grand Forks, Odegard had a meteor by the tail. Though he didn't realize it at the time—or, given his capacity for instant grandiose extrapolation, maybe he did—he now held the secret combination to academic credibility, rapid growth, jet airplanes, hundreds more students, international exploits, gee whiz technology to turn his critics greener with envy, and much more. Because, on that short hop between Bismarck and Grand Forks, it dawned on John Odegard that he had a rich uncle named Sam. One of the great pork barrel stories of all time then unfolded at the hands of Odegard and Pat Hurley, his new found insta-bond North Dakota buddy from the BuRec.

During that flight to Grand Forks, Hurley asked Odegard if he thought his aviation program could help him find a role for North Dakota to play in the HIPLEX project. It was like asking a used car salesman, "Do you think I should buy a car?" Odegard began machine-gunning ideas.

"We connected right away," says Hurley. "He was very outgoing, very enthusiastic. Everything just fell into place. Click, click, click."

The click they decided on actually made sense, in a robbing from Miles City to pay Paul kind of logic. Since the HIPLEX program was supposed to measure the downwind effects of cloud seeding, and since North Dakota was downwind of Miles City, who better to study those effects than...*someone in North Dakota!* Hurley didn't worry at all that Odegard's operation had no scientists on staff, had no way to collect scientific data, had no computers to analyze the scientific data they couldn't collect, and had no people who knew anything about computers or research or weather.

Mere details. Besides, Odegard seemed so sure that these minor speed bumps could be worked out, that it seemed a done deal. The two men went to the hockey game but the real skating took place in their non-stop strategizing about how exactly they could make it all work.

Part of their plan involved that NSF education grant which had yet to be disbursed. Some of the $235,000 was earmarked to buy a weather radar which Odegard had vaguely planned to put on top of Gamble Hall.

Hurley reminded Odegard that the HIPLEX program required a mobile radar unit. Quickly Odegard connected his unpurchased NSF radar unit with the now unused portable control tower at the Grand Forks Airport. It was the same unit Les Severance had helped the city get through the FAA, and because of it, the city later got federal funds to install its permanent tower.

The mobile tower was now sitting unused at the airport. Odegard knew he could buy it as surplus for $1. He also knew the perfect man to find a way to put the radar unit and the control tower together with some wheels: the clever engineer who held the patent on the world's first cantilevered hangar—the jazz man, Don Smith.

Click, click, click.

Hurley loved the idea.

The morning after the hockey game in late December 1973, Hurley and Odegard went to see Tom Clifford. "I knew Tom," says Hurley. "I told him we wanted to get aviation involved in HIPLEX, but it might mean getting *you* involved. Weather modification was kind of a controversial thing. 'Don't fool with mother nature.' People were skeptical at the university of weather modification. So I said we might need someone of your stature to talk to local groups. Tom said he'd be glad to do that."

Hurley then returned to Denver and laid it out for his boss, Archie Kahan. The lack of UND's credentials troubled Kahan, so he flew to North Dakota to see for himself. According to Hurley he was impressed by the old school talent and responsibility of people like George Hammond and Don Smith. And, like so many others, he fell under the spell of Odegard's boyish enthusiasm—though he later confessed to Smith, "the only experience John's had with the atmosphere is flying in it."

Still he'd seen enough to entrust Odegard with a small contract. Kahan and Hurley arranged for the NSF to immediately transfer $50,000 of its education grant to the BuRec. It would send it on to the North Dakota Aeronautics Commission who would pass it along to John Odegard. The NSF also amended its grant to specify that the department of aviation combine the radar with the old control tower. The radar, it said, must be converted to a state-of-the-art unit producing data in brand new digital and not the tired old analog form.

Kahan was impressed at how quickly Don Smith devised a way not only to add a digitized radar unit to the old control tower, but how effortlessly he converted an old trailer into a custom built transport system for hauling the device.

At that point, UND had a mobile weather radar that fit the specs of the

Frank Argenziano, foreground, headed the aircraft maintenance operation for years, dating back to the early 70's. It was his idea to paint the floors of the hangars white, which played a role in landing a contract with Chinas Airlines. Don Smith (right) acted as Odegard's deputy and helped establish solid relationships with senior officials at China Airlines.

This old radar globe, being hoisted above the UND airport barn, became an essential part of Odegard's strategy to win weather modification grants from the Bureau of Reclamation. Those grants eventually led to the start of a Department of Atmospheric Sciences, broadening Odegard's aviation program into an aerospace endeavor.

HIPLEX downwind project exactly. In fact, nobody else on the planet had one so exact, meaning the BuRec could "sole-source" it. The HIPLEX contract wouldn't have to be put out for bid. It was then agreed at BuRec that UND submit a proposal for downwind research, and it would likely get the grant.

Not a moment too soon. The very next day, says Hurley, the Washington office of the BuRec officially announced its choice of Miles City, Montana; Goodland, Kansas; and Big Spring, Texas as the three HIPLEX research sites. North Dakota's senior Senator, Quentin Burdick was visiting the Commissioner of Reclamation in Washington at the time. He immediately demanded to know why Williston, North Dakota hadn't been selected. The commissioner nervously called Archie Kahan in Denver and asked him the same question. Without missing a beat, Kahan told him that North Dakota had just been selected as the downwind research site.

"It saved our ass," says Hurley. "It would have been embarrassing and difficult without it. It just fit in perfectly."

Burdick was satisfied, and though he was astute enough not to say anything like, "I didn't know they knew how to do that" Kahan later wrote a letter explaining his decision. Hurley recalls that Kahan essentially told Burdick that he could have found it very easy to say "no" to someone who had no meteorological or research background. But he'd been impressed with Odegard's enthusiasm and the capabilities of his department.

In February of 1974, the NSF grant came through at $274,700. But the following year the BuRec's downwind grant that Odegard received was for $1.4 million. It was certainly enough for Clifford to quash the idea that the department of aviation wasn't bringing in enough money. It effectively doused for good the white-hot issue of separating the airport from Odegard's control.

But if that were the sum total of the intricate plotting that went into this grant, it probably wouldn't be worth the telling. The *way* the $1.4 million was approved, however, repeated itself in other BuRec grant applications over the next decade. With the downwind grant, for example, Hurley was the one delegated by his superiors to write the BuRec's official justification to Congress of the importance of the HIPLEX program. In that report—written after he'd met Odegard—he was careful to repeatedly stress the necessity of 'downwind research' and 'mobile radar' and other items that, coincidentally, only UND could supply.

For the appropriations hearings, Congressional committees normally submit written questions in advance to various agencies about why they should fund a particular program. "They often don't know what questions to ask," says Hurley, "so they would call John at UND for advice on what questions to ask. John would call me. I would write out the questions and give them to John, who would give them to the Congressmen. They'd then send the questions to the commissioner of reclamation for answers. The commissioner would call us and ask us to write the answers. It fell to me. It was a matter of me writing the questions and the answers.

That's exactly what happened with a lot of things here in North Dakota. That's how the system works. Agencies don't admit it, but you've got to know how to play within that reality. Smaller states like North Dakota are more adept at it. They have to be."

How adept? Before the BuRec weather modification funding dried up in the mid-Eighties, Odegard had raked in $13 million in various spin-off weather-related grants. He would go on to make successful inroads in several other federal agencies as well to get more cash from that pipeline.

The interesting thing about a federal grant strategy is that if Odegard could do it, why didn't other schools on campus also take advantage of the way the system worked? Many of Odegard's campus critics never quite appreciated, even years after his death, that the millions he spent—on buildings, high tech equipment, modernization and expansion of his fleet, sophisticated simulators, expansion of airport facilities and creation of entire departments almost overnight—were specifically earmarked by federal legislation and didn't come by way of Tom Clifford's patronage. The standard complaint was, "Shouldn't that money be going toward a new library?" Yet it seems no one ever sought out the Pat Hurley of the Washington library bureaucracy.

All that was in the future. In the meantime, in 1974, all Odegard had to do was to figure out a way to get a team of qualified scientists and computer experts to give up their promising careers in the lower 47 and move to North Dakota to measure some clouds.

Oh, and while he was at it, he ought to get hold of a computer system bigger than anything UND had at the time or ever expected to have.

Mere details.

Student pilots at the Odegard School have access to this Canadair Regional Jet Flight Management System Trainer. Combined with sophisticated software, students are able to make an easy transition to jet training.

The modernistic Clifford Hall at the John D. Odegard School of Aerospace Sciences is named for the former president of UND, Tom Clifford, long a supporter of John Odegard. A walkway (right) connects to Odegard Hall.

CLIFFORD

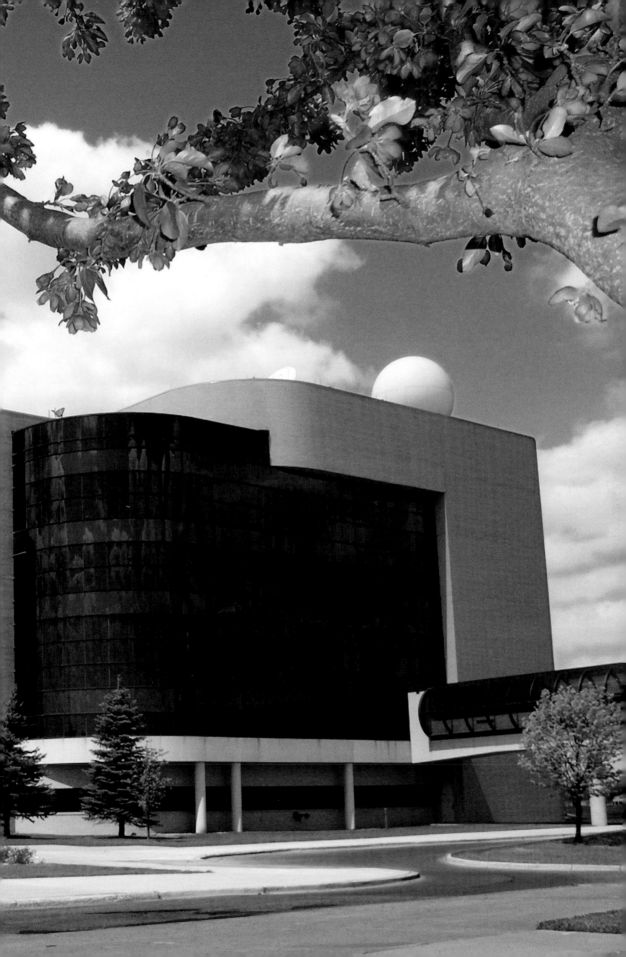

This $1 million walkway, connecting Ryan Hall with Clifford Hall is part of a network of connectors that link all of the buildings at the Odegard School. Founder John Odegard liked to call them gerbil tubes.

The fully modern airport facility at the Odegard School dwarfs the Grand Forks city airport. Forty years ago when John Odegard started the school, the only structure that existed in the university's section was the rounded Quonset hut structure, top right. Above it are the parking lot of the main city airport building and the city's runways.

Odegard Hall (at top right with radar dome) was the first building erected in 1983 at what then was called the Center for Aerospace Sciences or CAS. In this current computer-rendering of the main academic campus of today's John D. Odegard School of Aerospace Sciences, Odegard Hall is flanked by Streibel Hall, far right. A walkway links them to Clifford Hall (center), which connects across 42nd Street to Ryan Hall, far left.

This sophisticated Air Traffic Control laboratory at UND prepares students for immediate entry upon graduation into jobs with the FAA. The Odegard School offers more ATC training devices than any other college in the world. The school also operates 225- and 360-degree tower simulators along with a 32 position radar simulator.

NASA Astronaut Selection

- Identification of people who are best and have a high level of functional balance and resistance to stress

- Psychiatric interviews & multiple psychological testing

- Monitoring during training and mission simulations for tolerance, ability to self-regulate mental, somatic and autonomic functions, endurance, interactive styles in group situations and ability to deal with uncertainty, fear and crises (externally/internally precipitated)

Lessons taught in this specially equipped distance learning classroom at the Odegard School are transmitted to students all over the world.

Among the cutting edge craft in the growing UND fleet at the Odegard School are several single engine Cirrus SR-20's (top, left). The Duluth-based Cirrus company manufactures some of its wing components in Grand Forks. One of the mainstays of the UND fleet is the Piper Warrior (top right). Thanks to the 100,000 plus student training hours flown annually from the UND facility, the Grand Forks airport has become one the busiest in the world.

Even during overcast conditions the ceilings in North Dakota are usually high enough for students such as Cyril Wrabec, left, and instructors to safely comply with Visual Flight Rules (VFR) requirements. Odegard students such as Jennifer Sawislak, (lower right) begin flying with instructors right away, drawing on a large fleet of university planes, most of them from Piper (top, right).

ODEGARD

Odegard Hall, formerly known as CAS I, was the first of several modern new buildings built on an old baseball field on the campus of the University of North Dakota in Grand Forks, starting in 1983. From the start, all of the buildings, with their radar domes, and elevated walkways impressed everyone with their Buck Rogers aura.

The best way out
is always through.
Robert Frost

Chapter 13
He Kept Beating on It 'Til It Gave

Before he met Pat Hurley, John Odegard had done a fair bit of traveling to conferences and meetings across the country, always looking for the right person or opportunity to boost his program. Certainly he traveled more than any other faculty member at UND. In those years, to save money, Gerry Skogley generally prohibited faculty or staff from airline travel. It was a time when even President Tom Clifford was technically required to seek permission of the governor to leave the state. Odegard, however, had his own plane—or his department's plane—and could fly wherever and whenever he wanted. In 1974, after the BuRec grant was approved, his travels took on a turbo-charged intensity as he shook bush after bush for scientists, computer wizards and more sources of funding.

Odegard was fearless about pitching his school to those people he really wanted. "He wasn't threatened by highly intelligent, superior people," says his old college friend, Lloyd (Corky) Everson, MD. "John was a master salesman," adds Cessna's Russ Watson. "He did everything right, and he didn't leaved any stone unturned. John realized very early that it doesn't make any difference what business you're in, relationships are important."

Yet to develop one set of relationships, Odegard had to skimp on others, perhaps explaining some of those feelings of guilt cited earlier by Pam Kvidt.

"When I got there in 1971, we had 50 students," remembers Jean Haley. "Four years later, we had 200 students and 15 to 20 planes. At first we were all used to the personal touch. John was always around when we had a problem. Then, he was always busy, or he was gone. He was out looking for funding, visiting politicians, rubbing elbows with everybody. He didn't have time to be our buddy or counselor anymore. Some of us felt a little miffed."

Yet with all of the upheaval of Odegard's comings and goings, his son, John Jr., and daughter Stephanie have no memories of their father ever missing important moments.

"My perception," says Stephanie, "is that even though he did travel a lot, it seemed as though he was always there. He always came to my basketball games. I remember a Parent's Night in high school. They were announcing everybody, and I remember saying, 'Where is he?' And he flies in at the last minute. He'd always say, 'I am just on time.' He always got there."

Meanwhile, even as Diane Odegard was coming to grips with her husband's trips, she also realized that when he was home and they took visiting faculty recruits to dinner, she was expected to carefully speak the company line.

"One time we were downtown at a restaurant with a couple," she recalls. "I don't remember who they were. I don't think they ever took the job. We were talking to them about a big position. We were getting into North Dakota and Grand Forks and what it was like. And I made the statement, 'Well there basically isn't any spring. We just jump from winter to summer.' Driving home that night John said, 'Don't tell them there's no spring. We're trying to get these people to come here and live here.' He was so upset with me for saying that."

Recruiting, however, wasn't the only reason for Odegard's frequent absences from the school. One very big reason was a folksy North Dakota farmer named Bryce Streibel who soon became another flying buttress to Odegard's dreams.

The two men had first met in the fall of 1970 when Streibel brought his college-aged son, Kent—an aspiring pilot—to Grand Forks to have a look. Streibel wasn't just an ordinary parent on a college run. First, he was a UND alumnus and an old college friend of Tom Clifford. After World War II, Streibel had tried his hand at such varied careers as design-engineering and the undertaking business. He eventually settled on his family farm in Fessenden, in central North Dakota. It was from there, starting in 1956, that Streibel launched a 32-year political career, first as a member of the State House of Representatives and later as a State Senator. In 1970 he was one of the most powerful men in the state, serving as the majority leader in the Republican controlled House.

Not surprisingly, Odegard rolled out the red carpet for Streibel and his son. The farmer-legislator was impressed. "He just exuded class and motivation that was unbelievable," says Streibel. "We developed a terrific rapport."

Streibel retired from the House in 1973. In 1974, he gave up the farming business which had made him wealthy. In between political offices, (he would go on to a lengthy career in the state senate) he was looking for a cause on which to spend his clout and energy. He settled on Odegard's program for two reasons. Number one, his son Kent was enrolled there as a student, and had impressed his father with his zeal.

"The second reason is selfish, probably," says Streibel. "I've always had the corny feeling that when the good Lord created us, he gave us talents and said,

'Okay, use 'em to your optimum and leave your mark.' I thought what better way
to leave my mark. Without John being the type of person he was, I wouldn't have
been as motivated. He was exuberant, ambitious, impetuous, a driving force. And I
really did want to see this thing go professional, so it would get national renown,
and so the kids who graduated from here would make *their* mark."

In his view, Odegard *needed* help. Streibel was aware of the ongoing faculty
unrest over the department of aviation. Not only had Odegard's critics tried to
separate him from the airport, they still wanted to separate him from Grand Forks
and UND. Many had begun orchestrating a move to have the legislature transfer
Odegard's program to the state vocational college at Wahpeton, south of Fargo.
Their tune was familiar: pilots should not be trained at an academic institution.

So, on April 15, 1974, Streibel paid another visit to Odegard and gave him his
assessment of the school's future. "I said, 'John, you have two options: move your
airplanes and students and faculty to Wahpeton. Or go professional. Create a
Center for Aerospace Sciences with your own building.'"

Odegard brought up the Bureau of Reclamation's new interest in his
department, but Streibel brushed it aside. "The BuRec is not going to be there
forever," he told him. "And this aviation department in Gamble Hall—I don't know
how you got them all packed in there—that's not going to be there forever. We've
got to go to Washington and open some funding doors."

So that's what they did. On their first trip, they spent three intense days going
from one meeting to the next with officials from any agency that had anything to
do with flying or weather modification. Many of those doors were opened and
meetings arranged by North Dakota Senator Milton Young. But Streibel also relied
heavily on an old pal from the fifties when both were young national
committeemen of the Republican Party. This was Mark Andrews whose family, like
Streibel's, had farmed central North Dakota acres since forever. At the time,
Andrews was North Dakota's solitary member of the U.S. House of Representatives.

Streibel made his goal clear to everyone he met in Washington: He wanted
Odegard's program to get a building of its own and to become more than just a
department hidden away in the basement of Gamble Hall, even though, in reality,
it was on the first floor.

Streibel and Odegard returned for a second round of visits to Washington, and
between 1974 and 1982, Streibel went back again eight times on his own nickel.
Each time, he would stop at the office of Mark Andrews who usually faked a groan
and said, "You again!"

In the meantime, Streibel met regularly with Tom Clifford to keep him
apprised. Clifford knew of the desire to create a center with its own building. He
warned Streibel to be careful, in his Washington dealings, about committing to
something beyond the university's ability to deliver.

By this time, Streibel had been appointed to a coveted and influential seat on

All three of these men, (left to right) Mike Poellot, Leon Osborne and Cedric "Tony" Grainger, played a key pioneering role in expanding the scope of John Odegard's department of aviation. Their work, along with that of former UND scientist, Pat Brady, led to the establishment of the Department of Atmospheric Sciences.

*to John
with best wishes,
John Burdick, U.S.S.*

The late U.S. Senator from North Dakota, Quentin Burdick, (second from right) helped channel federal funds toward Odegard's fledgling aviation department. In this shot from the mid 70s, Grand Forks Mayor Mike Polovitz, Odegard and an identified local businessman welcome Burdick to an event in his honor.

the State Board of Higher Education. Things were looking mighty good for him and his crusade until 1976. In that year, Sen. Milton Young died, and with him went much of that door-opening, favor-doing arm-twisting power built up over a long career in the Senate. But it was the same year that most of Streibel's high-placed contacts in the Washington bureaucracy—people who owed their jobs to either Richard Nixon or Gerald Ford—vanished with the election of Jimmy Carter.

With Democrats now in power, Streibel figured he was back at square one. His only choice was to somehow make like a Democrat. He and Odegard crafted a clever strategy. Odegard nominated Streibel to be a member of the newly formed National Weather Modification Advisory Board, a position sure to open doors and show Streibel where the money was.

Streibel, however, was one of 362 people to be nominated for the 17 available slots. He had one hope. He had served in the North Dakota House with Tom Stallman, a Democrat, who was now the administrative assistant to North Dakota's other U.S. Senator, Quentin Burdick. It was the perfect connection, and in 1977, Streibel got his seat on the board.

However, he found out to his discomfort at the very first meeting of the board that each of the other 16 members held the highest scientific, technological, medical or academic pedigrees. At that meeting, they went around the room introducing themselves to the group. As the Ph.D.s rolled off one tongue after another, Streibel grew less and less confident of being able to make much headway with such heavyweights. When it came his turn to speak, though, he pulled a full North Dakota. "What the hell can I say?" he blurted. "I'm just an old Wells County farmer."

As it turned out, none of those 16 big shots was a farmer—the very type of person weather modification was supposed to be helping.

"That made a hit with them," says Streibel. "From then on it was, 'Yeah, yeah, yeah, the old Wells County farmer.' If you think that didn't open doors for me and my project… Holy Toledo these people knew people. They didn't know anything about the aviation department at UND but pretty soon they knew I was trying to get a building."

While Streibel was off sowing corn in Washington, Odegard was continuing his hunt for scientists. He'd done well right away by recruiting Lee Bruney, a man described as a computer genius, and Pat Brady, a Ph.D. from the University of Oklahoma. Brady was exactly the type Odegard wanted: a two-for-one hire; he was a meteorologist but he also was proficient in computer science. One of Brady's first jobs was to organize and analyze the radar data from the downwind field studies.

In these early days of the BuRec HIPLEX project, Odegard's team was collecting its radar data and storing it on computers physically located in Denver but connected through telephone lines to Gamble Hall. The BuRec grant didn't cover costs for UND to buy its own sophisticated computer. Odegard's initial requests for assistance from the university's computer center were met with glazed eyes. They didn't have the resources, they didn't have the time, they really didn't want to deal with Johnny's boys. It was that kind of thinking that spurred Odegard—who seldom took no for answer—to decide to simply create his own computer department within a department. His staff became so successful at handling the technology that a few years later Tom Clifford eventually transferred the university's entire computer science department out of the School of Arts and Sciences and into Odegard's control.

But all that is jumping the gun. In 1976, it became apparent to Odegard that something more efficient had to be done regarding the computer situation. That year, he hired another bright scientist, a man he'd bumped into briefly out in New Town, North Dakota. Odegard had flown out there with Pat Brady and the BuRec's Pat Hurley for a tour of one of the state's cloud seeding projects. Their guide for the two hour visit was a young man named Mike Poellot, an interesting combination of a private pilot and meteorologist. He was also familiar with radar and computer operations. Odegard was impressed with his guide and filed Poellot's name away. He would soon be needing it.

At about that time in the summer of 1976, Pat Hurley had told Odegard that the BuRec wasn't interested in doing all of the processing of data being collected downwind at the four different HIPLEX sites—including the one operated by UND. The question asked in Denver was who could and would do it?

Hurley suggested to Odegard that UND become the central processing site for the data. The data, however, was stored on 12-inch reels of magnetic tape. Without their own computer, it would mean flying UND scientists back and forth to Denver to do the work.

Hurley remembers Odegard lamenting his lack of a computer then quickly switching mental gears. "He snapped himself out of it," recalls Hurley, "and said he would get one. Nothing was impossible to John."

Odegard approached a local automobile dealer named Bob Absey whom he'd met during his ongoing courting of Grand Forks merchants. Absey, who also owned the Ramada Inn in Grand Forks, was a private pilot and liked what Odegard was doing. So much so, that in April, 1977, he bought a $100,000 computer and leased it back to the department of aviation. Overnight, Odegard had his computer department. Absey's monster machine was somehow squeezed into the Gamble Hall offices. It was bigger and faster than any computer the university had. Its state-of-the-art half-megabyte of memory—laughably small today—astounded Odegard's critics and did severe damage to the joints of their noses.

To go with the computer, Odegard directed Pat Brady to hire that nice young man Mike Poellot, to help process the BuRec weather data. In all, about 10,000 data tapes were processed by Odegard's team. The new computer made the work very routine, but not without one minor controversy. Each of those 12-inch tapes was mailed by the BuRec to UND. There it was processed then repacked in its bulky canister to be mailed back to Denver. To haul those canisters back and forth from Gamble Hall to the university post office, Odegard bought a small, red, Radio-Flyer wagon. Eventually all of that hauling and the weight of the tapes wore out the wagon's wheels. Odegard, with tongue in cheek, sent the BuRec a bill for the $25 cost of new Radio Flyer wheels.

Since federal bureaucrats are required by law to check their sense of humor at the door, they were not at all amused when the bill for wheels on a little red wagon landed. "They went, 'What the hell are we doing buying wheels for a wagon?'" recalls Poellot. It became, literally, a federal case. Higher-ups in Washington were consulted and after a good bit of buzz, buzz, buzz, someone, at some point broke ranks with a hint of a smile.

In the meantime, even though the computer made the data job possible, it still required an enormous amount of work. That's what ultimately led to the hiring of programmers and computer technicians and the start of Odegard's own scientific computing center. One of those hired was Leon Osborne, a highly regarded graduate student in meteorology at the University of Oklahoma. The story of how he was recruited by Odegard falls into the classic category.

Osborne was finishing his master's degree with every intent of staying in Oklahoma to earn his doctorate. A faculty member told him he ought to apply for a programming position at UND—not to actually take the job, but for the experience a job interview would give him. And, of course, the free trip to beautiful, downtown Grand Forks.

"I had never heard of Grand Forks," says Osborne. "I asked friends who'd been there what it was like. Someone said they do have trees and they are somewhat civilized." Another friend indicated that he kept hearing drums beating when he left work at night in western North Dakota. He said the drums beating for hours on end made him nervous.

Completely unenthused, and a bit fearful, Osborne flew up to North Dakota in the fall of 1978. He was shown around campus and quickly got the impression he was but one in a long line of people coming through for an interview. After a day and a half he was bored, unimpressed, and anxious to get back home. The last item on the agenda was a 10-minute talk with John Odegard.

"I'll never forget it," says Osborne. "He had a tiny office, but he greeted me with a firm handshake and a big smile. I instantly felt, 'this is a nice guy.' Everyone else I met, it was almost like they were going through the motions. He asked me to tell about myself. I had high expectations of where I wanted to go in my life. John just sat there and smiled and listened and nodded his head and let me finish. He had a way of leaning back in his chair and taking everything in.

"When I got finished, he leaned forward onto his desk and made a statement that changed my life. He said, 'That is very impressive. If you come to UND, then anything you want to do, I will support you. Whatever you want to do, you can do it here. As long as you're successful, I'll support you.'"

The conversation turned into a brainstorming session about what Osborne could do at UND. The 10 minutes stretched into 45. When he finally got to the Grand Forks airport, Osborne called his wife in Oklahoma and said, "We're coming to North Dakota."

Though Kathy Osborne was supportive, her parents, Osborne's parents and his advisors at Oklahoma were aghast. All tried many times to talk him out of the move, pressing the point that North Dakota meant the end of his career. It was a school, they said, known in the atmospheric sciences world for nothing but cloud seeding—something his faculty advisors considered the lowliest rung on the meteorologist's ladder.

"But John so captivated me," says Osborne. "There were times when all he'd have to do was give me a friendly smile and a slap on the back and I was good for another 80 hours that week."

By 1979, Odegard had one more captivation in his first round of raids on the scientific world. Someone at the BuRec told him he should talk to a man named Cedric "Tony" Grainger. With a Ph.D. in atmospheric sciences, Grainger had been eking out a living as a professor of meteorology at the state university in Oswego, N.Y. It was a small, Lake Ontario town whose heyday had come and gone during the War of 1812. Yearning for something a little more Twentieth Century, Grainger hired on with a research company in Boston. But he never actually made it to The Hub. His new company—which had just won a contract with the Bureau of Reclamation's HIPLEX project—shipped Grainger immediately to Goodland, Kansas, a city much like Oswego but without the lake. It was there, as director of the Central States regional HIPLEX site, that Grainger met John Odegard.

Grainger's initial reaction to a job offer by Odegard was absolutely "no." North Dakota looked even less like Boston than Goodland. But as with Osborne and so many others, Grainger was soon drowned in endless waves of Odegardian charm and fervor.

"John could sell a freezer to an Eskimo," says Grainger.

His one final doubt was the salary structure at UND. Leaving Kansas for North Dakota would have meant a significant salary cut, and Grainger told Odegard he just couldn't do it.

"I thought that was it. Case closed," says Grainger. "But he went out and managed to raise the entire salary structure."

The problem was that even though Odegard's faculty were being paid out of a federal grant—known in academia as 'soft money'—the university insisted they comply with the same low salary structure as all other UND faculty.

"But," says Grainger, "John knew where to hit 'em where it hurt."

Odegard told the rest of the university faculty if he couldn't hire the people he needed, he would no longer be landing all of those federal grants. Without them, the department of aviation would collapse. As much as his critics hoped for that, he pointed out one small drawback: A portion of every federal grant is reserved by the university for its own use. These so-called "overhead funds" are eventually spread across the campus to various departments and colleges according to a formula. It had always been a sore point with Odegard that his department couldn't recoup all of its grant dollars. But now he turned it to his advantage. Shut me down, he said to his critics, and you'll lose your share of that money. But, if you let me set my own salary structure, you literally have *nothing to lose*.

Maddeningly, it made some sense to them. Thus Odegard could not only hire Grainger, he could and did raise the salaries of everyone else in the department.

"That was the last argument I had for not coming," says Grainger. "John would not say die. He kept beating on it until it gave. I've never met anyone who could do that to the degree John could. So, I thought I could live in North Dakota for a few years if I had to. Here I still am 27 years later."

There is no hope
for the satisfied man.
Frederick Bonfils

Chapter 14
Rogue's Gallery

John Odegard has been variously described as a man constantly on the move, a man who moved fast, a man who sometimes moved too fast, a man who couldn't sit still, a blur of nervous energy, non-stop, mile-a-minute, go, go, go. He drove cars fast—sometimes 80 miles an hour down DeMers boulevard in Grand Forks. He drove boats fast—even his family's houseboat—fast enough sometimes to knock passengers off their seats. Of course, he always pushed to the limit the speed of whatever airplane he was flying. And ever since he'd learned to fly, John Odegard had wanted to fly faster. He would talk often about wanting to try a jet. When he started his department, the dream went from trying to get a jet to actually buying one.

"Once we got research money," says Don Smith, "even though we were not breaking even, we had cash. When you have cash, the entire image of your program changes."

The image, Odegard decided, could only be enhanced by adding a jet to the fleet. But Pat Hurley, supervising the BuRec grant, talked him out of an immediate jet purchase. "I told him we can't afford it. You have to start off with something more connected to cloud seeding. What he needed was a high performance turbo prop, a Piper Cheyenne, for instance."

Hurley then helped him get separate funding from the BuRec and the renamed State Weather Modification Board to buy a used Piper Cheyenne turboprop, equipped with cloud seeders and data collection equipment. While it satisfied the BuRec's requirements, it didn't keep Odegard's stubborn mind off a real jet. He argued that by owning a jet, the department of aviation would stand a better chance of getting more funding for weather modification, since a jet could fly higher and faster. In the back of that stubborn mind lay the idea that if the school had a jet, he would be able to fly it here and there and back again whenever he wanted. Tony Grainger says that all of the scientists in the department knew that Odegard had no real interest in weather. "The interest was in the jet, there was never any question about that."

In 1979, the Bureau of Reclamation awarded Odegard a continuation of his HIPLEX downwind grant, this time for $2.7 million. Like a boy running to the candy store clutching a brand new nickel, Odegard went jet shopping. He wanted a Learjet because Learjets came not only with a high performance wing design but with an image sustaining prestige. For comparison, Odegard also tested a Cessna Citation II. In the end the Citation seemed a better jet for the program—oddly, because it didn't go as fast as a Lear.

"We didn't want to go zipping through clouds so fast that our instruments didn't respond," says Grainger who'd been hired specifically to help outfit the new jet. "The Citation was a good compromise."

The used Citation they bought was owned by the Mercedes-Benz Corporation. Odegard himself went to Germany to take possession and to fly it back to North Dakota. Grainger then heavily modified the Citation for use as a research platform. After adding $200,000 worth of equipment, it would have been a tight squeeze to ferry passengers in the jet, but that became a moot point. Odegard was disappointed to learn that the BuRec required the Citation be dedicated totally to atmospheric research.

The lease-to-buy agreement that made the deal possible, allowed UND to own the Citation after 10 years—120 monthly payments of nearly $25,000 each. To house the new Citation and the Cheyenne and other new aircraft, Odegard spent more money erecting new hangars and consolidating space at the airport facility. That also meant hiring more mechanics and support staff.

Making all of those payments meant Odegard's program had to keep bringing in the funding. More than ever the department of aviation took on the look of a business. Odegard was very clear in telling his new scientific wing that being successful in whatever they wanted to work on meant generating enough funding not only to pay their own salary but to keep the ever expanding school aloft. So the fund generating began in earnest. Or, as dispatcher Jerry Nelson puts it, "That's when things really started to move."

In 1920, Luigi Pirandello, an Italian novelist and playwright who was awarded The Nobel Prize, introduced a play called "Six Characters in Search of an Author." It tells of a group of characters whose lives are unfulfilled because their playwright cannot finish their story. It may be a stretch—and if so, surely a harmless one—but by 1979, the year of the jet, John Odegard had fashioned something akin to an unresolved Pirandello. He had two high performance planes in a new hangar. He had hired Tony Grainger, Pat Brady and Mike Poellot to work on the HIPLEX

Mike Poellot was a private pilot and a meteorologist when John
Odegard met him in the mid 70s. He was also familiar with radar and
computer operations, the perfect hire when Odegard landed Bureau of
Reclamation grants to study wind effects. Now, both Poellot and his
atmospheric science colleague Leon Osborne have received the
University of North Dakota's highest form of academic recognition as
Chester Fritz distinguished professors. Poellot later helped start the
Department of Atmospheric Sciences, which he now chairs.

project, and Leon Osborne to manage the processing of weather radar data for the Sierra Cooperative Pilot Project (SCCP)—a wintertime companion project to HIPLEX.

Still, Odegard had fired up all of them to believe they could do anything they wanted. Thus played out "Four Scientists and Two Really Cool Airplanes in Search of a Department." Unlike Pirandello's story, this one had a happy ending.

Their story properly begins in 1974, the year Don Smith designed that mobile radar tower using funds from the National Science Foundation. Pat Brady had gone after the NSF grant and, using the mobile tower, designed the modest research effort—the observations in support of cloud seeding research and operation—that launched Odegard officially into the meteorology business. It was during that period that Odegard started his training program for weather modification pilots. Because his scientific faculty was small, Odegard began using guest lecturers from various universities and agencies such as the National Oceanic and Atmospheric Administration (NOAA) and the BuRec. He videotaped those lectures on cloud seeding theory and its methods and recycled them for years.

These were the aviation department's first "academic" classes, that is classes not strictly related to aviation or aviation business. Because of the hiring of his own well-credentialed faculty, Odegard was eventually able to cut back on guest lecturers.

Mike Poellot took over the ad hoc science group, still formally under the department of aviation. Before long he and his colleagues noted that the university course catalogue for the department of geography listed several classes in the atmospheric sciences that weren't currently being taught. Odegard's faculty proposed a joint program with geography whereby they would teach meteorology classes—not to student pilots, but to regular science students on campus. This rankled the geographers as blatant turf infringement, and they refused.

"In 1978 when I arrived," says Leon Osborne, "there was already open warfare between the two departments." Most of the campus, he says, supported the geography department, and viewed Odegard as "a rogue outsider." Osborne remembers difficulty in convincing the department of geography that he or any of Odegard's science faculty knew anything. "We were seen as illegitimate," he recalls, "and there were times it got hot."

Legitimacy for Odegard's growing faculty of scientists would come, but not for many years. When it did, it was the strength of their research on those BuRec contracts, along with their own independent projects, that began to sway opinion. In the late seventies and into the eighties, however, their research got more respect statewide and nationally than it did in Grand Forks. Of course, the Odegard scientists did have a great advantage over other schools: They were the only aviation department in the country with an atmospheric sciences research division and its own research jet. More and more, says Tony Grainger, his colleagues branched out into their own airborne research projects. The Cheyenne was used for state cloud-seeding project, and the BuRec gave the department a weather research grant that took the Citation to California for a time. The Citation jet was constantly in use for one research project or another.

At the same time, many of the scientists in the so-called atmospheric sciences research division missed the classroom and yearned to take on more teaching duties. In 1979 and 1980, the research projects become so numerous that the science faculty—as in most science programs in the country—found itself in need of full-time science students to help them with field research. But there were no students at UND dedicated to a full-time meteorology program. And trying to recruit researchers from elsewhere to North Dakota—without high profile name recognition—flew like a brick.

"We thought if we had a meteorology program, we could hire our own students," says Grainger.

But if the earlier idea of a joint geography-aviation meteorology class didn't go over, the proposal of a joint meteorology degree program with geography was bitterly resisted. Many particularly resented that the idea was officially proposed by Bryce Streibel—who was then a member of the state Board of Higher Education, *the* ruling body in upper academe. At a meeting of the university's curriculum committee, some anti-Odegard faculty members talked about a Streibel conflict-of-interest. Their comments were reported in *The Dakota Student* in September of 1980.

They demanded to know why Odegard had brought in Streibel—an unprecedented intrusion, they felt, for a state board member. Pat Brady, Odegard's academic chief for the entire department, told the committee that Streibel hadn't been brought in, but had taken the action on his own. It appears that remark convinced few. Brady also noted that any Odegard faculty who taught meteorology would be paid for out of federal grants.

That didn't seem to sink in with critics. English professors complained that President Clifford had once established that "service courses" such as, well, English, were to be a priority. They should be the ones to get the money for new programs, not Odegard.

According to the *Dakota Student* article, William Dando, the chair of the department of geography, said, "he thought he'd be the director of such a program. 'I have the Ph.D., and I have the experience...I entered into this willingly and in good spirit. But there's a problem working with Mr. Odegard."

Leon Osborne confirms that. "Sometimes John would say things at a meeting with geography then leave and send me in. It was bitter."

As for that curriculum meeting, Nancy Krier, the author of the *Dakota Student* article, reported that the geography chair had earlier told her that his department "required a Ph.D. to be an instructor. 'John Odegard has none.'" That thought was echoed to the reporter by Bernard O'Kelly, dean of the College of Arts and Sciences.

"O'Kelly…says later there's a misunderstanding. The college requires the highest degree possible in a field. 'No one in the world that I know of gives a Ph.D. in aviation.'"

To someone like Tony Grainger, who held a Ph.D. in atmospheric sciences, it was a telling and familiar comment.

"People looked at flying airplanes as being something not really up to their level," he says. "There were a lot of people who just plain hated the idea of the program. For a long, long time we were seen as just part of the airport group. For a long time I was listed at meetings as 'Tony Grainger, airport.' We were just out there in never-never land. It was an uphill struggle."

The issue came to a head at a second curriculum committee meeting a week later, also covered by Nancy Krier of the *Dakota Student*. William Dando complained to the committee that he'd tried to get a meteorology program approved several years earlier but was told there was no money in the university budget. Of course, the money *wasn't* in the university budget; it was in Odegard's budget because he'd gone out and got it for a specific purpose—with the help of people like Mark Andrews and Bryce Streibel, of course.

Streibel himself showed up at that second meeting and made a pitch for the program, citing "a crying need" for meteorology. And yes, he told the group, a meteorology program had been proposed earlier and he had supported it. But the state Board of Higher Education had turned it down. Now he was a member of that state board, and he meant to change the weather.

At that point he left no doubt about his intent. According to Krier, Streibel said that if the program wasn't approved by the faculty's curriculum committee he would be "discouraged and disappointed," and that his fellow board members would be "a little perturbed."

After that, the program passed. Odegard's upstart, quasi-department within a department began teaching non-aviation students meteorology through a degree program in conjunction with the department of geography. Three years later, Odegard negotiated the complete transfer of meteorology out of the geography department along with the right to teach all of the academic courses that fell within the atmospheric science discipline.

But not before he cut one of his famous corners on two wheels.

The devil's deal he'd made with the university to get what would become the makings of a new department, was to surrender half of all overhead monies that he received in grants. That money would be sprinkled through the university infield to assuage hard feelings.

However, according to Tony Grainger the arrangement left the budding science group with less than adequate resources to gear up and run a full department of atmospheric studies. But Go-Go-Odegard was anxious to get it started, figuring once more that once he demonstrated success he'd be able to talk his way out of the deal. In fact, it would be another 20 years, long after Odegard's death, before the department of atmospheric sciences began to receive its own sprinkle of state appropriated dollars through normal university channels.

Incidentally, even though the department of aviation was now in the meteorology business, not until the mid-eighties did the stubborn university bureaucracy see fit to grant those scientists department status.

"We would ask for department status," says Grainger, "and we'd be sent to the Faculty Senate, who then sent us to the curriculum committee. But we already had a curriculum in place. They would send us to plant services and so on. Eventually, after many years, people across campus found out 'Gee, they have scientists over there doing research on interesting things.'"

Yet the critics were fighting a losing battle. It's true, many would remain opposed to aviation until they retired, died or left for more pure levels of academe. However, forces greater than their disgust were at work now. A critical mass was building for Odegard, and not just in the matter of his scientists seizing control of the atmosphere. It happened in an area more important than mere air.

That, of course, would be money. The money first began to flow with the BuRec grant. But in 1979, it started to pour in from an unexpected source.

By early 1979, having completed his tenth year of windmill tilting, Odegard resembled the description of a James Bond martini—shaken, but not stirred. To note the accomplishment, the *Minneapolis Tribune* sent reporter Dick Youngblood to Grand Forks for a look-back profile of the program. By now the department had 42 aircraft in its fleet and a budget approaching $2 million. Odegard had added new undergraduate career paths to his curriculum. Students now could major in aviation administration, aimed at jobs in the business end of the industry; airport administration, which trained them in all aspects from business operations, to flight operations; aeronautical studies, which allowed them to major in any arts or science program at the university, along with earning an instrument-qualified commercial pilot's license.

In the article, Odegard bloviated on the growth potential of the airline and general aviation industries. Youngblood described him as "an engaging blend of guile and gall," and attributed his success to "'an end run' he had pulled around stodgy bureaucrats."

That story, picked up by wire services, was widely read in Minneapolis and throughout Minnesota, but also in dozens of towns and cities across the Upper and Lower Midwest. High school students who identified with Odegard's engaging rogue image realized that they, too, wanted to fly. They soon flooded the university with applications. Within two years, the enrollment of aviation majors had jumped from a little more than 200 to 500. In another two years, it stood just shy of 700—the majority of the increases said to be a result of that one article.

It was becoming harder and harder to ignore the lowly department of aviation now that it had become the third largest program on campus. But the critical mass was still coming together. Bryce Streibel was still out there campaigning for a building and promoting something he called a Center for Aerospace Sciences. Mark Andrews, meanwhile, elected to the U.S. Senate in 1980 and proud of the *Time* magazine article that labeled him "a pork barrel expert," was still backing him.

Andrews wasn't the only politician whose ascendancy to higher office helped John Odegard. Ronald Reagan took office as President of the United States in 1980 and proved not only tough enough to withstand an assassin's bullet but a bitter strike by the Professional Air Traffic Controller's union. In August of 1981, about three-fourths of the nation's 17,000 federal air traffic controllers went out on strike. Reagan gave them 48 hours to get back to work or lose their jobs—something no one but Reagan ever thought he'd really do. Within a week, 9 out of 10 strikers had been fired, leaving gaping holes in the functioning of airport towers and control rooms across the country. This came only two years after federal deregulation of the airline industry made competition for customers a frenzied affair. In short, falling ticket prices meant air travel in 1981 was steadily increasing. Leaders of the controller's union thought that gave them the leverage for getting a better contract. They were wrong.

This incident, say most of Odegard's friends and supporters, triggered one of *the* classic moments in the life of a man who could think 500 miles ahead of everyone else. A duck pouncing on a June-bug had nothing on the opportunity-seizing speed of John Odegard.

"John was always looking for the silver lining in any cloud," says Don Dubuque. "When the ATC strike came on, he said we need to be training air traffic controllers." Critical mass now had been reached and a chain reaction was about to level the playing field.

One important link in that chain had been forged ten years earlier when Odegard and Gary Kiteley took over the University Aviation Association. Because he and Kiteley had led the way to curriculum standards in the mid-70s for aviation schools, the UAA had taken on a certain gravitas in the industry. The UAA had moved to Auburn, Alabama where Kiteley took over as its executive director.

When the Federal Aviation Administration realized it needed to replenish the nation's supply of air traffic controllers, Odegard was ready and waiting. It wasn't that the FAA stumbled across the UAA and saw in it a reputable organization that could help it devise a training program. It already knew John Odegard quite well, from the Les Severance days to the constant nudges from Bryce Streibel and Senator Mark Andrews.

The FAA knew Odegard so well that it had tried to lure him away from UND. He'd even been to the FAA headquarters in Washington and seen the office he'd occupy if he took them up on their job offer.

So it was natural for the FAA to turn to the UAA for help. At the FAA's request, Kiteley put together a blue ribbon committee of educators chaired by Odegard. They developed a curriculum called Airways Science, designed to train students in the air traffic control industry and to become a feeder system for future FAA employment. Working behind the scenes, Mark Andrews, chair of the Senate Subcommittee on Transportation Appropriations, got the FAA to kick in more than $60 million to get the curriculum inaugurated at 66 colleges—one of them, UND.

"It was a major revitalizer of collegiate aviation education," says Kiteley. "It brought many small schools, including minority institutions onto the collegiate aviation map for the first time. UND probably benefited the most, but John's philosophy was 'Let's help others too.'"

The money was to be used for equipment, facilities and buildings. Construction soon began on Odegard's very own building on university land that had once been the UND baseball diamond. The old Wells County farmer down in Fessenden was smiling. Streibel shares the credit with Andrews, who credits two additional people: Odegard's reputation for showing results with grant money— which made it easier, politically, to get him more federal funding. And Tom Clifford's subtle, remote-control shepherding of his protégé.

"It's a tribute to Tom Clifford what happened and what worked and how it was done," says Andrews. "We all worked together because we were all North Dakotans. We wanted to see something like that succeed. And John was enthusiastic enough to transfer it to the staff he had."

Leon Osborne was one of those who thrived on the boss's enthusiasm. Odegard liked to call him Willard—after Willard Scott, the goofy TV weatherman. Willard Scott was someone no real meteorologist wanted to be compared to, but Osborne felt honored to have an official Odegard nickname, and he never complained.

"John and I would sit in his office at night and vision," says Osborne. "Cooking up schemes on where we wanted to go."

Osborne had noted that the weather radar system the school had purchased long ago for the HIPLEX project was sitting unused and in disrepair at the airport. He proposed retrofitting the unit into a Doppler radar. Doppler was so new at the time that it wasn't even cutting-edge yet. Not even the National Weather service had deployed Doppler units. But on the back of an envelope, Osborne scratched out a budget for building the Doppler and more.

"I gave it to John and said this is how much money we need," Osborne recalls. "But we also needed to start creating a place where students could come and look at weather. John had certain things he wanted. He felt strongly that people needed to see a bit of the flash. I gave him a funding request for computer graphics equipment. It was brand new, particularly color graphics. I said we also needed an instrumented facility, a weather facility. I said if we could do this we can create a research program around it."

Odegard liked the idea and passed it to Mark Andrews who included a budgetary earmark in the next round of federal appropriations. It fell together as if it had been designed that way from the start: scientists looking for meaning, a meteorology program looking for space, an atmospheric sciences department looking for viability, two former farmers and one former baseball field looking for a building, a jet-propelled department chair looking for credibility and a big, soft sack of folding money.

"We wouldn't even be sitting here if it hadn't been for that money," says Kent Lovelace today. "John saw the air traffic controller's strike as an opportunity. Others saw it only as an inconvenience."

A footnote: The Airways Science program led to the UAA establishing a council on aviation accreditation which, today, is the certifying body for the nation's aviation education programs. UND's was one of the first four programs accredited. However, an Inspector General's report effectively killed the Airways Science project in the early nineties for gross inefficiency. It noted that the FAA had spent more than $100 million over 14 years on the project. While most graduates of the program went on to attractive jobs in the aviation industry, the lumbering FAA bureaucracy managed to snag only 43 of them.

Odegard's escape from Gamble Hall began in June 1982, when construction started on that new building—an $8 million, 55,000 square foot structure with bells, whistles and looking-good style. Besides classroom and office space, it would contain a $330,000 atmospherium, a type of planetarium with an indoor amphitheater seating 160. Another feature Odegard added was a science-fictionish spherical "radome" on top of the building. It contained weather radar for use by meteorology and weather modification pilot classes.

Through his life, Odegard liked to convey the impression that he knew a little bit about just about anything. From lecturing a friend on the proper way to barbecue a salmon (before he'd ever barbecued one himself) to offering Ph.D.-like advice on building a fire by the lake, he had that maddening knack for consistently being right when he had no right to be. The design of that first building—there would be three others within ten years, courtesy of Uncle Sam—was no different. Even before the structure was designed, say its architects, Odegard had in mind a complex of buildings, all connected by walkways over high traffic roads.

"I still have the drawings we did of all the buildings," says Grand Forks architect Bill Schoen. "John always knew he was going to go across Forty-second Street. He always knew he'd need a collection of buildings to take him where he wanted to go. He always had the big picture in mind."

More than once, he and his partner Jim Kobetsky walked into Odegard's office to see him waving his arms in excitement, telling them he wanted this or that on his building. "Sometimes we looked at each other and said, 'How are we going to do that?' But John was a visionary, always pushing the envelope. I could kick myself for saying this, because I'm awfully proud of this community. But my first

The architects who designed CAS I took their inspiration from John Odegard's creative, visionary personality. Many marveled at the Buck Rogers-like appearance of this building, in such stark contrast to the stately traditional architecture of the rest of the UND campus.

The strike by air traffic controllers in 1982 gave Odegard the idea to start an ATC training program. The FAA helped fund the first building of what became the Center for Aerospace Sciences or CAS. CAS I (right) was soon joined by CAS II (left) when the university's computer sciences department was shifted to Odegard's program.

impression of John was 'Wow, what's this guy doing in Grand Forks?'"

Kobetsky says that Odegard was so fired-up and hands-on about those buildings that they were designed to reflect his personality. When the first building was dedicated in August of 1983, it was different by far from any building on campus, in town, in North Dakota or anywhere between Minneapolis and Seattle. It was a mysterious blend of steel and black glass topped off with that golfball-like radome, giving it a futuristic Star Wars look.

"I think the buildings look like him," says Kobetsky. "They have an aerospace feel. A feel of what was in his mind, all the futuristic stuff, the good and the goofy ideas. He was always into that latest and greatest of Buck Rogers."

When the first building opened as the Center for Aerospace Sciences, or CAS, one might have assumed that with such an in-your-face fulfillment of a dream John Odegard would at last be satisfied. Typically, Odegard wanted more. He'd badgered Clifford for months to make his department an aerospace college. He wanted very badly to be a dean. But Clifford told him to slow down.

"Making it a center established a different identity from a department," says Clifford. "It had all the benefits of a college: a person could go into an aviation-arts curriculum or a business curriculum or an atmospheric science curriculum. John wanted to go right to a college, but you have to be cautious about doing that. I'd already established a center for education on campus. I viewed a 'center' as transitional. John didn't appreciate that a lot. He resisted it. But he was fine in the end. He had the luxury of no choice."

Chapter 15
Typical Odegard

The meteorology victory was still being savored in early 1981, the start of a decade that would produce many more victories and generate a grudging respectability for Odegard and his once humble program. But not before the kind of tragedy that any pilot or flying program dreads.

On February 17th of that year, a freshman aviation major named Dwight Widseth, from the small town of Crystal, Minnesota, embarked on the second solo flight of his young career. According to George Hammond, Widseth took off from the Grand Forks airport in a single engine plane and was circling just east of the airfield. Hammond and John Odegard at that moment were five miles away, inbound from Colorado in the new, twin-engine Cheyenne turboprop.

According to Hammond, the tower operator told Widseth to fly straight across the tower and then downwind. This he did, coming around and entering the traffic pattern for other student aircraft just taking off. Meanwhile, a second plane, carrying an instructor and a student pilot who was working on his commercial pilot's certificate, was just taking off into that traffic zone. Theirs was a high wing airplane, meaning they couldn't see Widseth's craft coming across and above their flight path.

Meanwhile, Odegard, piloting the Cheyenne, had just contacted the tower for permission to land. The tense reply from the tower stunned him. It said, "We just had a mid air."

The student pilot and his instructor had flown directly up and into Widseth's plane without ever having seen it. Their plane managed to make a forced-landing and both occupants walked away. Widseth was not as fortunate. He was killed when his plane crashed at the end of the north-south runway.

"We got on the ground as quickly as we could," recalls Hammond. "John went right out. We were all very distressed about it. I'd seen similar things in World War II, returning aircraft that didn't quite make it. You don't forget those kinds of things."

Odegard certainly didn't. It was the first fatality in the history of the program, and it hit him hard from both a personal and professional standpoint.

"It was a black day here," says Diane Odegard. It got blacker and sadder when Widseth's only surviving parent, his mother, arrived. They learned that her late husband also had been killed in a flying accident while serving in the Air Force.

"John became very, very close with Dwight's mother, Jeannie," says Diane. "We had her with us almost every parent's day for many years. She was like a member of our family."

The FAA conducted a full scale investigation. Some thought the accident had been caused by miscommunications with or from the tower, some thought by pilot error. As a result, a new, more modern tower was built, and when it was dedicated to Dwight Widseth, his mother was there on Odegard's arm.

"John was so wonderful with her family," says Diane. "He took her under his wing. She came to our banquets, and John made sure her life here at UND was good even though she had lost her son." A memorial quilt, hand stitched by Widseth's mother, still hangs in the dispatch area at the airport.

The incident was especially jarring because serious accidents had been rare in the program.

"Statistically we should have lost many students," says Bob Reis. In fact, based on the general aviation industry's average fatality rate, the 400,000 flight training hours flown between 1969 and 1981 would have generated two per year, or a total of 24. UND had but one in that time span. Over the next 26 years—and 2 million more flight training hours—it has been The Odegard School's only fatality.

"Accidents were tough on John," says Reis. "He took them very personally, as if we pushed too hard or didn't do something right. He wanted to understand the cause and circumstance to put us into a preventive mode for the future."

Safety already had been heavily stressed in the program. Jean Haley remembers to this day the lesson she was taught by Don Smith in the very first class she took.

"He described how accidents happen," she says. "He said every accident results from a chain of events. There is something that is not quite right and something compounds it that also is not quite right. Then another thing happens. He said at the end of that chain you're going to crash, and you probably will die. However if at any point in that chain you recognize it and break the link, the accident doesn't happen. His point was to teach us to recognize those links. He was so right with that analogy. It was the single most beneficial bit of wisdom I've heard in my career."

The program experienced a second fatality two years later, this time involving the brand new helicopter program. With permission of the flight staff, a student named Wayne Twitero had flown a UND helicopter to his home in Sisseton, South

Dakota. But shortly after he took off to return to Grand Forks, he ran into heavy fog. An airplane can fly by instruments through fog but a basic-training helicopter cannot. Twitero was killed when his copter crashed close to the Sisseton airport. To date, in 40 years of flying, those have been the school's only flight-training fatalities.

The helicopter program, by the way, was another of those eighties-era victories. It originated through an Odegard deal with Hughes helicopters and became a mainstay for training U.S. Army ROTC cadets enrolled at UND. The program started with one helicopter and five students. By 1984, there were 21 students and three leased Hughes model 300C gas powered helicopters in the fleet. Unsatisfied, as usual, Odegard took out a $460,000 loan from a Denver bank to buy a Hughes jet turbine powered chopper. As he had so long ago with his Grand Forks bank, he got the loan based on projected student fees. He charged $400 an hour for flight time. Each of those 21 ROTC students needed 50 hours flying time to be certified by the FAA. Those fees came directly from the U.S. Army, under a program Odegard titled, somewhat romantically, "Air Battle Captain."

By then, things were rolling along in high gear. American Airlines had heard about Odegard's program and asked him to help train flight engineers for FAA certification. The flight engineer is the most junior member of an airline cockpit crew. In large airliners they serve as second officers, and in smaller craft they are often the co-pilot. American already had a 6-week flight engineer training course in Dallas, but the cost of using its 727-simulators for that handful of trainees in the final two-weeks of the program was breaking the bank.

American agreed to fill in its classes in the simulator phase of its program with UND students who completed a 3-credit, one semester course that Odegard's staff devised. Essentially those students could skip the first four weeks of the Dallas training and then maximize the use of the American simulators. An added benefit for both sides: the plan created a pool of FAA certified flight engineers ready for employment.

The FAA added another feather to Odegard's plumage when his department moved into its new digs and transformed itself into a center. The agency granted the new Center for Aerospace Sciences full examining authority. Odegard students could now be officially FAA-certified as pilots by UND flight staff, instead of waiting for an FAA examiner to come to Grand Forks to test them.

Odegard himself, the man who once ran afoul of the rules of the sky, became a certified FAA examiner. The agency had been impressed that 98 per cent of the 525 aviation students who gradated from UND's private and commercial piloting courses between 1981 and 1983 had passed their final flight check on their first try. Because of that, the FAA took an unprecedented step of relaxing the minimum number of flight hours a UND student was required to log before qualifying for a commercial license.

According to a 1983 article in the official magazine of Northwest Airlines, "The

FAA has stated that UND's quality of training far exceeds any quality of experience measured by total hours." And speaking of total hours, the magazine noted that in 1982, the Grand Forks airport was ranked the fifteenth busiest general aviation airport in the world, with 85 per cent of its business and 95 per cent of its tower operations coming from UND. Today the volume of traffic at Grand Forks has made it the world's ninth busiest general aviation operation.

Meanwhile in 1983, when the aviation department became the Center for Aerospace Sciences, it not only began the official takeover of meteorology and atmospheric sciences from the department of geography, it stole away the entire department of computer sciences from the College of Arts and Sciences. According to Grainger, Tom Clifford had become concerned, with the dawning of the high technology era, that the computer science department wasn't moving fast enough. "He wanted John to make that a credible program," says Grainger. "So he said, 'Take it.' He had a lot of faith in him."

It was not lost on Odegard's critics that this internal shakeup meant that the computer sciences department arrived at the Center for Aerospace Sciences with its budget and salaries fully appropriated by the state and university. Of course, the salaries of Odegard's weather faculty were paid for out of grants and aviation faculty out of student flight fees. In the stroke of a pen, Clifford had given Odegard not only an empire but respectability. He now had a sizable chunk of state funding just like every other college in the university.

To accommodate all those new computers and professors, a second CAS building was put on the drawing board. Streibel and Andrews went back on the scent of those oh-so-soft millions. All of this sudden respectability and real estate was only enhanced by the progress being made by Odegard's atmospheric scientists. With more than $6 million in federal funds already filtered through their various downwind and cloud seeding research efforts, they could report an average 10 per cent increase in precipitation across North Dakota. Professor Pat Brady said it was responsible for hundreds of millions of dollars in increased crop yield.

In an article in the *Grand Forks Herald* at about that time, reporter Chuck Haga summarized the success of Odegard's program. He tallied the 750 majors, the 200 staff and the 60 aircraft—which he referred to as "The UND Air Force"—and noted that Odegard had been trying to sell his weather modification services to Prince Hasan bin Talal of Jordan.

And yet, said Haga, Odegard bristled at the suggestion he was a wheeler-dealer. "I don't like that," he told Haga. "We're considered mavericks here and that hurts us a lot. Aviation is a dynamic business, a cash-sensitive, technical business. You've got to be prepared to run with things. If we blow an engine on a twin-engine aircraft, we've got to air-freight a $28,000 engine in here the next day."

He couldn't leave it at that without further fanning the flames of war. "The university community isn't used to that," he said. "They're used to pondering for six months whether to buy a typewriter."

Odegard had bigger things on his mind than typewriters. He'd told Haga he was unhappy that faculty hadn't availed themselves of the air service his center offered them. "The university should be using our services much more," he said. "We have a plane that goes to Bismarck nearly every day. The seats on that plane should be filled."

They weren't always filled, however. Except for a handful of doctors and administrators—who had successfully used the flights in their state-wide campaign to promote the expansion of the medical school from a two-year to a four-year model—the service that Odegard had envisioned in his master's thesis wasn't flying very high. Haga reported that "suspicion and bad publicity...have made some university personnel nervous about taking advantage of UND aircraft."

No one was suspicious that the planes or pilots were unsafe. Other than just not wanting to contribute to Odegard's success, some of the more practical concerns of the faculty emerge in a story told by Don Larson. He is the computer technology chief at the university's School of Medicine and was a long time friend of Odegard's. He says Odegard called him one day and offered to fly him to Bismarck for an official meeting. Larson said fine and flew to his meeting and back, a happy man. A few days later he was stunned when he got a bill for the flight. He eventually chuckled it off to the "typical Odegard" syndrome—the little matter of not being told up front that the flight wasn't free.

The bad publicity, however, was a much more involved matter and was the one spot of trouble Odegard found himself in with the FAA in the first part of the decade. It came about because of a bill his staff shouldn't have sent.

The root of the problem lay in Odegard's dissatisfaction with having a Citation jet that he couldn't fly very often because of its limited research mission. To rectify that he went to two reliable Grand Forks supporters and essentially invented an airline.

The first visit was to insurance man Dave Vaaler who agreed to help underwrite the creation of a small North Dakota airline. Odegard's selling point, says Vaaler: "He said no commercial airline flies east and west in this part of the country. If you want to go to Bismarck you have to fly to Minneapolis and then to Bismarck. So he got us kind of interested in it. The interesting part of it is that John stayed out of it from a financial standpoint."

Instead, he played matchmaker. Odegard flew Vaaler to Minneapolis and the two of them, "walked right into the office of the president of Northwest Airlines and sat down, and John told him what we wanted," says Vaaler. "The president, Joe Lapensky, said I think we can fix you up. He got four people from various parts of the organization, brought 'em in and says to us, 'You guys come back at 4 p.m. this afternoon; I'll have this all set up for you.' I was impressed. I was sitting there talking to the president of Northwest airlines. John just walked right in there. We came back at 4 p.m., and he says here's what we can do for you."

Long story short: Vaaler and some investors ended up forming a company called Northern Airways and it eventually fizzled. It would have flown passengers from Devil's Lake east to Grand Forks. Northwest then would have flown them wherever they wanted to go. But a competitor beat Vaaler's group to the Devil's Lake franchise. Northern and its small fleet floundered for a few years trying to drum up business.

"It got to be a little bit of a strain financially," says Vaaler. "So we finally bellied it up. To be honest, I don't think anybody ever researched whether it would be profitable. That was John's idea, and I thought it was a good idea. We found there wasn't that much movement between Grand Forks—East or West—which is obviously why the airlines didn't do it. John didn't really factor in the financial costs. I said I don't know how the heck you pay for it, and we didn't, until we found out we were the ones who proved it couldn't be done. I like to think of that as a service and not just a plain dumb thing to do."

But before it bellied up, Northern played a role in getting Odegard a second jet. Besides putting the arm on Vaaler, he'd gone to friend, supporter and potato mogul Tom Ryan. He asked Ryan to buy another Citation for the school. Ryan and the McCormick family of Fargo bought the jet, according to Chuck Haga in the *Grand Forks Herald* in late 1981. The jet was then leased to Northern Airways which operated it under an FAA Air Taxi License. Northern then turned around and sub-leased it to the University of North Dakota for student training.

Odegard used the new Citation to fly state and university personnel on public business, according to *The Herald*. He also contracted back with Northern to fly private charters and emergency medical flights, using the seven people on his staff—including himself—who'd qualified to fly the Citation. Students usually served as co-pilots. For this service, Northern paid UND $35 an hour. Under FAA regulations this qualified as a "limited service" and didn't require Odegard to maintain a separate air taxi license.

So far, so good. A short time later, in a well-publicized incident, Odegard used the new Citation to fly a young burn victim from Fargo to a hospital in St. Paul, Minnesota. In the midst of accolades for such service, a group of fixed base operators complained. They noted that there was no way they could compete for such emergency or public service flights, not at the unusually low rate of $35 per hour that Northern was charging. In essence, they said, the state—through the university—was subsidizing a private company in unfair competition against them.

It was a familiar argument, raised way back when Odegard first started his program. In the *Grand Forks Herald* story, Odegard recited an equally familiar defense: He was providing an invaluable educational opportunity for students and, in the case of the burn victim, "a valuable community service."

That might have been the end of it, until a complaint was lodged with the FAA.

Dave Vaaler, left, a Grand Forks insurance executive, was the original insurer of the airplanes in the UND Flying Club, and, for many years, those in John Odegard's aviation program. Odegard also encouraged him to buy several small planes and lease them back to the university. Vaaler and Odegard often took short hunting or ski trips with friends. Ardel (Casey) Vilandre is at right. (Man second from left is unidentified.)

North Dakota Governor Ed Schafer (left), a friend and staunch supporter of John Odegard's aerospace program, invited him during the 90's to the world famous "One Shot Antelope Hunt." Held annually in Lander, Wyoming for celebrity sportsmen, Odegard, was joined by Tom Ryan (second from right), the millionaire Grand Forks potato grower who supported many of his initiatives, and Charles "Bud" Jacobi (right).

It seems that Odegard had flown a group of doctors and administrators from United Hospital in Grand Forks to Minneapolis to demonstrate the jet's capabilities. Nothing wrong there, but somehow a bill was sent by Odegard's office to United for the cost of that flight. That meant he was technically operating a transport-for-hire without a license.

It was a small, probably harmless technicality, but nobody was chuckling this time over another "typical Odegard." The Chicago office of the FAA found that Odegard had been in violation. He protested that it was much ado over nothing, but the incident played large and ugly in the news, and there were some who enjoyed seeing the Master Salesman in even lukewarm water. The incident came to a close when the Master Incident Closer, Tom Clifford, wrote to United Hospital and to the FAA. According to the *Herald*, he said the bill had been sent by mistake and that Odegard would never do it again.

And, after he emerged again from the Clifford woodshed, he never did.

There's just too much to see
waiting in front of me
And I know that I
just can't go wrong
Jimmy Buffet
Changes in Latitude

Chapter 16
What's So Attractive About North Dakota?

It's hard to talk about power without mentioning money. Those with the one usually want or get the other and then want more of the first. John Odegard, naturally, was different. He was certainly a powerful man whose influence brought in a lot of money—which only enhanced his power. Yet, though it almost certainly could have, that power and outside funding did little to enhance his personal wealth.

Obviously, Odegard was not a poor man. He had a modest home in a desirable section of Grand Forks and a modest lake cabin in Minnesota. He had access to any plane he wanted, and if the one he wanted wasn't in his fleet, he often found a way and a rationale for the university or someone else to buy it. That's how he got a series of float planes that he could use to train students in the techniques of water takeoffs and landings. He could, for instance, fly students to his lake cabin for training. Or he could fly himself and family or friends to the cabin for a weekend of paperwork.

His daughter Stephanie remembers flying back from the lake cabin with her father one Sunday evening. Sunburned and sore from a weekend of swimming and water-skiing the pair took off from Pike Bay near Minnesota's Cass Lake in a Cessna 185 floatplane owned by the university.

Though he'd never pressured his children to learn how to fly, Stephanie's brother, John Jr., had earned his private license on his sixteenth birthday. "But I never took to flying," she says. "I was more interested in cars. I always wanted to get in his lap when he was driving."

That night her father spent most of the flight to Grand Forks showing her the ins and outs of flying. At one point he allowed her to take the controls.

"He guided me towards the setting sun, explaining different functions of the plane," she recalls. "He liked to focus on the exciting capabilities of the plane, like simulating zero gravity and taking advantage of the clear skies to fly at as low an altitude as possible. We strained to spot running deer and farmers finishing weekend chores in the fields, while keeping a constant eye on the skies around us for other aircraft. I saw the sparkle in his eye, that pure sense of comfort, excitement and passion. That plane was filled with love. If it could be measured in pounds, we would not have been able to lift off."

At the end of the flight, Odegard credited Stephanie with her first two hours of flight time, signing them into a log book he'd started for her. But that would be the extent of her piloting days. Her father never pressured her to do more. "He was amazingly encouraging," she says. "Just non-specific. He let everyone of us know we could accomplish anything."

Certainly, he was a walking, flying, living proof of that philosophy. And though he was never personally wealthy, he always sported that million dollar look and gave the impression that the cost of just about any item was a mere bag of shells. He'd mastered the dangerous yet quite useful art of using other people's money to accomplish just about anything.

For instance, in the grand manner he would often picked up the check at a fancy restaurant. A standard gesture in the business world, but not in the sedate world of academe. Often on the morning after entertaining VIPs, Odegard would call Earl Strinden, the powerful state legislator who also headed the UND Alumni Association.

"It was not unusual of a morning that John would call and say he was going to come over with a number of bills that he needed help getting paid," says Strinden. "He had entertained someone here or there. Maybe he took some FAA people skiing or had others up to Oak Lake." [This was a remote, fishing resort in Canada, accessible only by float plane, where Odegard frequently took VIPs—from airline presidents, to U.S. Senators.] Strinden had a discretionary fund that he often used to help various programs, but he admits that Odegard got the bulk of it over the years.

"There was a reason," he says. "I saw John doing things that other deans weren't doing. He was much more an entrepreneur, someone who understood marketing and how to build contacts outside the university. I really felt this was a very good investment. I felt too many in academia were cloistered. They did not truly understand the real world of competition and the need for initiative and entrepreneurship. I felt they were too comfortable."

Even so, Strinden admits that sometimes he had to shake his head over the frequency at which Odegard took pains to show that he was a non-cloistered, uncomfortable, entrepreneur.

"John would call me and say he had some bill coming due, some dinners at the Ramada, a fishing trip, this and that. I'd say, 'How much is it John?' He'd say, 'Oh,

Though Stephanie Odegard chose not to pursue a flying career, she remembers fondly a flight alone with her father as the two returned to North Dakota from the family's lake cabin in Minnesota. Her father let her take the controls and guided her in a magical moment she cannot forget.

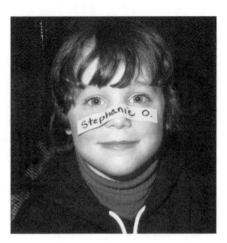

Though he never pressured his children to become pilots, Odegard's son, John Jr., took early to flying. Here, his father congratulates him on the occasion of John Junior's 16th birthday and his first solo flight. Odegard, Jr., graduated from his father's program and is now a vice-president with NetJets.

God, it's about $850 dollars.' I'd say, 'Well, send it over to Michelle.' And so some time later Michelle comes in with a whole handful of these bills and says, 'What do you want me to do with them?' I'd usually say, 'Get 'em out of my office. Just pay 'em. I don't want to look at 'em.'"

The fact is, with all of the perks of his power, Odegard didn't have any real money of his own. Not the kind he could have had if he'd taken any of his suitors up on their generous job offers. More than once, big aviation companies like Cessna came calling with job offers and enormous salaries. Once, in fact, his perceived archrival, Embry-Riddle University made a serious pass at him.

A dogged competitor, whether on the racquetball court or in a Brooks Brothers suit, Odegard had set his sights on besting Embry-Riddle from the moment in the late sixties that he started looking around and taking stock of *his* potential rivals. They were also UND's potential rivals, but to Odegard, competition was always personal.

In 1967, when he began going to the national air meets, several big name schools such as Ohio State, Purdue and the University of Illinois boasted aviation programs. Any one of them in those days was miles ahead of his own quasi-department. But in his eye, Embry-Riddle was the current best in the business. It traced its beginnings to a small airport in Cincinnati in 1925. According to the school, barnstormer John Paul Riddle and entrepreneur T. Higbee Embry began training some of the hundreds of young flying enthusiasts who bubbled up in the years following the first world war. Later known as the Embry-Riddle School of Aviation, it expanded to a campus in Florida where it trained 25,000 pilots and technicians for the military. During the Korean war, it trained maintenance crews for the Air Force.

In the mid-sixties, under President Jack R. Hunt, it consolidated its training in Daytona Beach, home today of its main campus of 5,000 students. It became a university in 1970, offering undergraduate degrees in aeronautical engineering, aviation administration and flight training. It has since opened a western campus in Prescott, Arizona, and changed its name to Embry Riddle Aeronautical University.

Whether Embry knew it or not, it became the Navy to Odegard's Army, the Yankees to his Red Sox, the Goliath to his David.

"Mention Embry-Riddle, and it would get him all fired up," says Bob Buley. "He viewed them as the ultimate rival."

Odegard was determined to have his school outpace Embry-Riddle in reputation and quality, in air meets and in equipment. In the early eighties, when Odegard got his jets, his atmospheric sciences department, his computer sciences department, his first building—with another on the way—he began to feel he had pulled ahead in the mythical national aviation derby.

In the summer of 1984, Jack Hunt, Embry's first president, was facing retirement. One of Embry's board members then was Darrol Schroeder, a North

Dakota native and two-star general in the Air Force reserve. Schroeder had known Odegard since forever. Both had been crop sprayers, each served simultaneous terms on the North Dakota Aeronautics Commission, and Schroeder was one of those invited experts who lectured Odegard's classes in the early days. Their friendship was cemented during the days of the late sixties when the fixed base operators were complaining of Odegard's unfair competition.

"I could see John's vision," says Schroeder, now retired. "I even met with some FBOs and said this is really not fair of you to be criticizing this, because it's aviation education, and in the long run it'll probably help you."

When Embry-Riddle began looking for a replacement for Jack Hunt, Schroeder turned the school's attention to North Dakota, simultaneously lobbying Odegard to take the job. At the behest of Embry-Riddle, John and Diane flew to Florida for a series of dinners and get acquainted meetings.

"Naturally," says Schroeder, "he blew them all away."

Embry's board of directors then flew to Grand Forks, had dinner at Odegard's home and began applying pressure for him to become their president. Odegard played his cards very close to his vest, so close that Diane didn't know he'd made up his mind until she read a story in the *Grand Forks Herald* that said he'd turned down the job.

He offered many reasons for staying. Too many exciting things going on in North Dakota. Unfinished goals. He didn't want to leave in the middle of Allen Olson's term as governor because they had such a good relationship. He said he didn't like it that Embry's academic mission wasn't as advanced as his program. Each reason sounded plausible, as did the idea that he was a North Dakota boy and just felt a loyalty to his home state.

No one, however, knew precisely what had changed his mind, not even the governor of Florida at the time, Bob Graham.

"I was at a governor's conference one day," says Allen Olson, "sitting across from Bob Graham. He looked at me and said, 'What's so attractive about North Dakota that I can't hire somebody?' I said, 'What are you talking about?'

Apparently Graham had been enlisted in the effort to recruit Odegard. According to Olson he said, 'I've tried to get the dean of your aerospace center to come down and take over Embry Riddle, but he won't do it. He won't leave North Dakota.'

"Bob just couldn't believe that North Dakota would be more attractive than Florida," says Olson. "And I said, 'Well, you don't know John Odegard.'"

Olson, however, felt he understood Odegard's decision to stay in North Dakota.

"John knew he was absolutely king of the mountain in Grand Forks," he says. "It was *his* aerospace center. He knew Embry was a small fish in a big pond and UND was a big fish in a small pond. John liked his friends and his freedom. He liked Grand Forks and didn't mind the weather."

Again, it makes sense. But still: a nagging thought that there was more to it.

"I've often wondered why John did things the way he did them and worked as hard as he did for the things he tried to get started," says Tony Grainger. "Why didn't he go on to bigger and better things?"

In the end, all of the above reasons were the answer to Grainger's question, but only as they came from the one source with the power to make Odegard listen. Simply put, Tom Clifford didn't want Odegard to leave.

That alone might have been enough, but Clifford sweetened the pot. He knew Odegard had lots of grand ideas for expanding his school, but he also felt the bureaucratic structure of the university with its intrusive state legislative oversight would slow the progress and frustrate his protégé. He proposed to Odegard the creation of an Aerospace Foundation, an independent arm that could flesh out those dreams and be free of outside obstructions. A foundation, he said, was the way to grow; it was the next logical phase of the challenge.

"John chose to stay in North Dakota, which was remarkable and a testament to commitment to what he was building there," says Byron Dorgan. "Others might have gone for the half-million dollars a year. That wasn't what John was about."

Mike Poellot makes it even simpler: "Without a doubt he could have made more money elsewhere, but this was his life."

And yet, Odegard's sister, Joanne Samuelson—as only a once-tormented and blackmailed sibling can—puts her finger on the pulse.

"John never did it for the accolades," she says. "He always had an agenda. Something in the works, some plan. It wasn't about winning. It was about the game. The winning would be anti-climactic to the game. With John, the game is the fun part."

One fun item on his agenda in 1984 was finding a prominent guest speaker for the school's annual spring awards banquet. Odegard told Earl Strinden he wanted to get Wally Schirra, one of the original seven astronauts. Strinden told him he knew of a UND grad in California who ran an oil company that employed the Apollo 11 astronaut Buzz Aldrin. He was the second man to walk on the moon after Neil Armstrong.

That grad was Cliff Enger, a 1934 UND law school alum. Prompted by Strinden, Enger got Aldrin to go along. The two flew out to North Dakota that spring. Even then, Odegard's agenda had an agenda. As soon as Enger and Aldrin landed in Grand Forks, Odegard packed them and Strinden into a university plane and flew them out to the small North Dakota town of Carpio, near Minot.

There was a ranch way out there belonging to a pair of wealthy old bachelor farmers named Paul and Oliver Skinningsrud. Both were pilots and had their own planes and a landing strip on their property. They'd given over their ranch for an air show staged by Odegard's old outfit, Pietsch Flying Service.

When Odegard and retinue dropped in—supposedly unannounced—on the surprisingly large crowd, Buzz Aldrin became an immediate hit. It was then that Earl Strinden found out that the promoters of the air show had sold tickets for their big dinner that night, advertising Buzz Aldrin as the featured speaker. No one had asked or told Buzz about it, and when he found out, he wasn't about to speak.

"He was miffed," says Strinden. "He said it wasn't a very polite thing to do. I said, 'Buzz, listen, these folks are so honored. This is the thrill of their life that you would be here.' I said, 'This was a mistake; we didn't ever promise you'd speak. Somebody kind of assumed that. You'd be doing all of us a great favor even if you just stand up and say hello.' Buzz thought about it and said okay he'd give a talk."

The dinner took place in an old Quonset hut, packed to overflowing.

"My God," says Strinden, "there were a lot of people out there in that little old community of Carpio, North Dakota. Well, I introduced Buzz, and he started talking. And so help me he really got on a roll. They shut down the buffet line when he started. There were a lot of people waiting to get their food. Buzz had all kind of things to talk about. It was fantastic. Buzz was brilliant. After 30 minutes, I went over and put my arm around his shoulder. I said, 'Say, Buzz, we got a heck of a problem here. We got folks who haven't gotten to the food line yet. I'm worried that if we don't let them go through that food line, they're going to beat the hell out of me, because I'm the guy who introduced you.'"

The next day Odegard flew the group up to Oak Lake in northwestern Ontario for some fishing. Strinden got so sunburned the first day that he decided not to go fishing on day two. Aldrin also stayed back in camp that day.

"We were sitting there, having breakfast," says Strinden, "and I just happened to ask Buzz a question about something. Well my God, he grabbed several napkins and started tracing things out and going on and on. After he'd filled up four napkins I said, 'For God's sake Buzz, do you think I can understand any of that? I'm still trying to figure out how a fly lands on the ceiling.'"

After many laughs, Odegard talked Aldrin—who'd earned a Ph.D. in astronautics from the Massachusetts Institute of Technology—into helping him create a space education program. Aldrin flew into Grand Forks four days a month for about three years, working with Odegard, Leon Osborne and others. It was at Aldrin's suggestion that Odegard persuaded scientific heavy hitter David Webb, a member of Ronald Reagan's Presidential Commission on Space, to come to North Dakota in 1987 and become founding chair of the space studies department. Today the department is the largest graduate program in the entire university.

Aldrin also served as a good will ambassador for the school and boosted its prestige. Bill Shea, the former FAA official who chaired the aviation department at the time, remembers he used to take Aldrin to lunch at Whiteys in East Grand Forks and on the way stop off a local elementary school and do a walk through.

"Buzz's job was to interface with faculty and students," says Shea. "Anytime he was in town, I'd have him in as many classes as I could. I wanted students to see and hear him, so they could say I had Buzz Aldrin in my class."

Typically, there was no money in the budget for Aldrin's modest stipend, but it didn't stop Odegard. He squeezed money for the space studies curriculum and Aldrin's costs out of the old standby: student flight fees.

"John enjoyed that little extra flair of bringing dignitaries to campus," says Leon Osborne. "John did everything with a flair."

But if there were moments people referred to as "typical Odegard," what about his mentor? Was there a "typical Clifford" action? The problem with applying the word "typical" to anything Clifford did is that it seems too basic, too hands on, too tactical. As Odegard was a dreamer and a plotter of plots, Clifford was one rung up on the stool, a strategist and a planner of campaigns. Not that Clifford was above the hands-on approach. He'd been national handball champion several times and while president, was easily the best player in North Dakota. But even in that highly tactical sport, Clifford focused not only on the ball, but on his opponent's moves and style—in other words, on his *typical* reaction to shots. His own classic moves came over the years to be characterized simply as *pure* Clifford.

Regarding the progress of the growing Center for Aerospace Sciences, there were several "pure Clifford" moments. One of them had to do with the funding of that second building, which came to be known as CAS II. It was the new home of the computer sciences department, and Clifford had relied on Bryce Streibel to work with Mark Andrews and other contacts to pay for it.

After Andrews got Congress to appropriate the $1.6 million needed for construction, Streibel went back to his Fessenden farm at the end of the day quite satisfied. But as the building was nearing completion in 1985, he got a call from Clifford. Streibel, a meticulous maker and keeper of notes, remembers the conversation well

"Tom said, 'Say Bryce, we don't have any equipment in there.'" Streibel almost fell over. "Well, I never heard of funds being appropriated for a building without equipment," he says. "But that's how it came in. Tom said we need another million."

Streibel, by now back in the state legislature—this time as a Senator—had enough clout to wangle a $1 million amendment to the regular UND appropriation bill. It meant he had to twist many arms and lay some prestige on the line. But he did it and went back to Fessenden at the end of the day satisfied.

A few weeks later, Clifford called him again to say he needed $500,000 to pave the parking lot of the new building. Streibel went through the ceiling. "Never in the history of North Dakota has a request been made for appropriating money for a parking lot," he said. "It's usually paid by user's fees."

John Odegard and Tom Ryan (foreground) were much alike: ambitious, industrious and successful. Both shared an appetite for hunting and fishing as well. Ryan was often a guest of Odegard's at his annual Oak Lake fishing retreat's.

But, Clifford explained, those fees would have to be paid by students who were already feed to the eyeballs with expensive flight costs.

"It was frustrating as hell," says Streibel. "You think that was an easy sell? My colleagues in the state house weren't very happy. I had to be candid and say, 'If you think I'm happy about this…' I testified in the House for the bill, and they asked me why they should fund it. I said, 'I can't answer that.' I think enough of them knew that I'd spent a hell of a lot of time and effort and my own money to get this whole CAS thing going."

In the end, the $500,000 was approved. Streibel called it a miracle. Others called it "pure Clifford."

But it wasn't the last miracle that year. For the first time ever, the University of North Dakota's Center for Aerospace Sciences won the National Intercollegiate Flying Association's championship. They were coached by a loyal Odegard recruit from St. Cloud (MN) State named Kent Lovelace. He'd started at UND in 1977 as a flight student and in 1980 was made an assistant chief flight instructor. In 1984, at Odegard's urging, Lovelace earned his master's degree and was promoted to assistant professor. With that promotion, Odegard also handed Lovelace the reins of the flying team. It made sense. In 1975, the last time the UND Flying Team lost in a regional final, it had been beaten by a St. Cloud team captained by one Kent Lovelace.

In 1984, his first year as UND's flying coach, the Lovelace squad finished second to Southern Illinois in the national finals. Afterward, a downcast Lovelace sought out Odegard and said, "I failed you. You'll have my letter of resignation when we get back."

Odegard frowned.

"Well I'm not going to accept it," he said, "so shut up."

Lovelace remained as coach, and his team won the national title the next five years running. In the 22 years since then, the Flying Team has won 14 national championships. Today Lovelace chairs the department of aviation.

> No great man ever complains
> of want of opportunity.
> Ralph Waldo Emerson

Chapter 17
We're Off on the Road to Morocco

Sometimes people see plots in every little thing. Most of the time, psychologists will tell you, those folks have simply sprung a leak in the tiny paranoia capsule we all receive at birth in case someday we really *are* being followed by men in dark coats and fedoras and need a little extra oomph to get the feet moving. It may explain why, in the summer of 1972 when George Hammond and Don Smith were threatening to quit, many Odegardians believed a plot was afoot by disgruntled faculty and gruntled bean counters to completely erase the aviation department from the face of the planet.

A hyper-extended imagination? Maybe, but only because there were so many people who actually did want the aviation department erased from the face of the planet. The tough thing about conspiracies is that very often one never knows for sure. There are, however, plots that make no bones of their existence, such as the one that emerged at just about the same time in those early seventies—but halfway around the world from Grand Forks on the continent of Africa. Their connection is sublime.

On August 16, 1972, Hassan II, the king of Morocco, was also feeling a bit paranoid. At the moment, he was aboard his private 727 returning from France to Rabat, his nation's capital. Only a year earlier, at a birthday party with 400 of his closest friends, he had survived a coup attempt. A battalion of his soldiers surrounded his ocean view palace and began firing. The 42-year-old king hid in a bathroom as 100 of his guests died in the slaughter. Afterward, Hassan took the hint and acceded to reforms, including a power-sharing experiment with an elected parliament.

Amazing how one little coup attempt can make you jumpy. In this case, Hassan II was quite wise to be a little paranoid. For outside the window of his jet on that August day in 1972, as he returned from France, a pair of Mirage fighters from his Royal Moroccan Air Force suddenly appeared. They immediately opened fire on his unarmed jet. With missiles, they blew out one of the plane's engines and with machine guns they drilled holes up and down the fuselage. It was then, according to royal legend, that the king himself got on the radio and, in a presumably unkingly voice, told the fighters they had killed the king. The fighters bought it—and would soon buy the farm as well. For by the time Hassan's plane landed, the conspirators had been rounded up, including their leader, the Defense Minister, who had until then been Hassan's right hand man. According to official reports, the minister took his own life rather than face the angry king. If he feared the angry king might personally put a bullet through his famous sunglasses—while they were still on his face—well, according to unofficial reports, it was one of those rare cases of justified paranoia.

Other unofficial reports said that it wasn't actually the king who got on the radio, at all. It was the quick-witted pilot of the royal 727 who shouted into his microphone "You idiots, you've just killed the king." Perhaps it's a loose translation, perhaps it's a loose anecdote. The fact is that the pilot, Captain Mohammed Kabaj, rode the gasping 727 to a safe landing and was quickly promoted to colonel by his grateful king. Later, he rose to general and command of the entire Royal Air Force.

Several years passed before these events would impact John Odegard. When they did, they had much to do with Morocco's geography. Once you get away from the country's temperate Atlantic coast region, Morocco's southern interior becomes dry and mostly desert. It is overshadowed, though, by the Atlas Mountains, a string of 11,000 foot peaks high enough to collect snow. There are even ski resorts in those mountains, not far from the dusty desert town of Marrakesh. Reservoirs up in those mountains collect snow to generate hydro-electric power and to irrigate farmlands.

In the mid-eighties, a drought held Morocco in its grip. As head of the Royal Air Force, it fell to General Kabaj to do something about getting the clouds that dropped snow on the Atlas Mountains to drop more of it. An amateur when it came to cloud seeding, Kabaj deployed his F1 Mirage fighters as weather modification platforms. But not only did the Mirage fly too fast through the clouds—firing silver iodide crystals through the armament that normally fired flares to outwit heat seeking missiles—it was an expensive plane to modify and to operate.

At about this time, a North Dakotan businessman named Dick McConn entered the story. McConn, a graduate of the United States Air Force Academy, headed a company, M International, that supplied various materials to the Royal Air Force. During a trip to Morocco he heard of the weather modification problems

One of the more fascinating characters who came in contact with John
Odegard was Mohammed Kabaj, the general who headed the Royal Air Force
of Morocco. Kabaj—here catching a fish during an Odegard outing—was
Odegard's entree into a contract to teach Moroccans how to get more
precipitation from their clouds.

vexing Kabaj. Not long after that, McConn visited his cousin, Chuck Muhs, at his cabin in Detroit Lakes, Minnesota—a favorite summer retreat for scores of North Dakotans. Chuck Muhs happened to be the brother of Bob Muhs, the loyal Odegard graduate working with Northwest Airlines. Because of his brother, Chuck Muhs knew of Odegard's weather modification work. He mentioned it to McConn.

Even as McConn was passing the word back to Kabaj in Morocco, Odegard and Clifford were well into plans to launch the independent UND Aerospace Foundation to take advantage of just such opportunities. The foundation would operate unlike the regular UND alumni foundation, whose principle mission was to raise funds to benefit university programs. The Aerospace Foundation was designed to generate the kind of business that could be run without the red tape of a university structure.

"It was designed to commercialize our product beyond the traditional student," says Bob Reis. "It was a business enterprise to create and privatize our intellectual property and our flight training capabilities."

Or, as Lyle Beiswenger—who had by now succeeded Gerry Skogley as vice president for finance—put it: "The Aerospace Foundation was created for the betterment and promotion of aerospace, period. This was not a fund-raising organization."

For a short time, the Aerospace Foundation existed under the umbrella of the UND Alumni Association. But the board of directors there felt they didn't have enough control over what John Odegard was doing or planning to do. When the Aerospace Foundation went out on its own, its board of directors was a tight-knit bunch: Tom Clifford, chair; John Odegard, president; and Lyle Beiswenger, treasurer. It had been created to do what an academic enterprise could not to: move at the speed of business, not the speed of academia. Because it was an independent body, it could okay business deals without going through any bidding process without having to clear any project with the state legislature or Board of Higher Education. It was also exempt from opening its books to any state authority.

Exploration of training and research contracts with foreign countries was a major goal of the foundation. Thus, when Chuck Muhs told Odegard that someone in Morocco needed more snow in its mountain reservoirs to water the flatlands below, he went immediately into high gear.

Kabaj and his family were invited to Grand Forks in the early eighties to hear Odegard's proposal for handling Morocco's weather modification. Odegard wasted no time whisking Kabaj, Muhs and McConn up to the Oak Lake fishing camp in Canada. Everyone got on famously and caught plenty of walleye.

Muhs later worked closely with Sen. Mark Andrews to get funding for the project through the United States Agency for International Development (USAID) and with Pat Hurley at the BuRec.

Kabaj went back to Morocco and laid the plan before his king. Hassan II liked the idea, and in short order Odegard, Tony Grainger and a few others were enroute to

Though the Moroccan contract never took off the way Odegard hoped, it was the first international project of the new UND Aerospace Foundation. Odegard and Tom Clifford were able to recruit aviation clients from around the world—clients who needed to train pilots—to supplement the income of the school and fund its growth.

When Odegard learned that Gen. Kabaj was the key to the Moroccan contract, he invited him and his family to America. He took Kabaj and several close associates up to Oak Lake, Odegard's favorite fishing retreat, and the two bonded.

Casablanca to nail down details.

The discussions went beyond the making of snow. Odegard raised the idea of UND training Moroccan pilots in the art and science of weather modification. He talked of conducting tests and research into the likelihood of super-cooled water existing in those Atlas mountain clouds—a pre-requisite for increasing precipitation.

Back in North Dakota, Odegard talked Governor Al Olson into returning to Morocco with him to lend his prestige and to help seal the deal with King Hassan II. Off they went, this time in an Air Morocco 747 on the king's dime, sharing the first class cabin and VIP status with several Moroccan ministers.

In Casablanca, they were put up in a 900-year-old palace that had been converted into a hotel—the same one Winston Churchill had stayed in during his World War II summit with Franklin D. Roosevelt and General Dwight D. Eisenhower. Olson took note of the carpet in the hotel. One of the two Royal Air Force majors escorting him said, "Governor, would you like one?" Olson replied, "Well, of course." He thought it was all a joke and didn't pay any more attention to it.

In Rabat, Olson and Odegard had their audience with the king. "He was diminutive," recalls Olson. "He looked like a tourist in Palm Springs. Western dress, very slight build, very un-king like. We gave him the North Dakota flag. He was very gracious."

After the deal was sealed, General Kabaj told Odegard that the Royal Air Force had just taken delivery of several new French Mirage fighters. Odegard insisted on flying one, though he'd never been in a Mirage in his life.

"Damned if he didn't get permission," says Olson. "He had a Moroccan pilot with him, but John insisted on being in the front seat. He had a flight suit on over his wall street investment advisor suit and his wingtips. I know he had a kind of flamboyant touch occasionally, but when he got in that aircraft, he was all business. He came back exhilarated."

They were then flown to Casablanca in a King Air for their return flight to America. Their Royal Air Moroc 747 was already on the tarmac, waiting for them. As they left the King Air, Olson noticed a truck speeding toward them across the tarmac. A rolled-up carpet protruded from the back.

"I'll be damned," thought Olson. Sure enough, it was his gift from King Hassan II, a beautiful carpet, woven by a tribe in the Atlas Mountains.

Back in the states, the Congressional delegation from North Dakota worked with the United States Agency for International Development (AID) to draft a $6 million block grant for weather modification pilot training in Morocco. All of it was earmarked for UND. The AID people got the BuRec people involved as a consultant, with Pat Hurley to oversee the project.

Meteorologist Cedric "Tony" Grainger, was hired by Odegard to equip a Cessna Citation Jet with instruments to conduct experiments in Morocco. Grainger stayed at the school and helped found the department of Atmospheric Sciences which he later chaired.

Everything looked good until a meeting in Washington in 1985 where higher politics scuttled the block grant idea. Instead it was chopped into several smaller grants, each to be put up for competitive bid. Hurley told Odegard the only grant among them that UND had a chance at was an $800,000 cloud physics program, involving the use of radar in weather research. There was already a good bit of interest in that one, and Hurley told Odegard his bid had to be financially tight. Odegard cut his proposal as much as possible but still had a huge chunk of money he'd need for fueling the Citation research jet. He called his new best friend, General Kabaj, and they struck a deal. Kabaj would have the Royal Air Force pick up the fuel tab. With that much knocked off the bottom line, UND won the grant.

Tony Grainger and his Citation research team spent two three-month periods in Morocco, one in winter and one in fall, measuring clouds for their suitability in cloud seeding. Before they started, however, they spent the better part of two years planning the project and doing both preliminary analysis of data and a final, full scale analysis.

"John had this view," says Grainger, "'Oh, it's going to be great, it's going to be great. We're going to go out and rent some villas by the ocean.' Of course it never was quite as great as John painted it to be. You sort of got this vision, listening to John, that there's going to be lots of spare time. But there was never any spare time. It ended up being a lot of work. It was always 'run as hard as you can.'"

And there were problems. Right away, the inertial navigation system in the Citation malfunctioned. It meant they wouldn't be able to tell how close to the mountains they were when flying through clouds. The Citation spent two weeks in London undergoing repairs.

Eventually, the Citation crew made their measurements and in preliminary analyses found some good cases for cloud seeding. But more trouble. Grainger and Odegard had been counting on getting more funding to do the ultimate final analyses of the data. The BuRec instead wanted a contract to analyze the UND data along with data from several other projects.

"We didn't really have the expertise to do all of it," says Grainger. "So we put in a bid with another company. They had us listed as a subcontractor for the aircraft data. But after all the bids were in, they decided to do it in-house. It left a bad taste in our mouths."

Though their time in Morocco didn't turn out the way Odegard expected, it was a success simply because it was the Aerospace Foundation's first international project and could be used as a calling card to future foreign deals.

It also marked the beginning of the end—as Bryce Streibel had once warned—of the flow of soft money from the Bureau of Reclamation. "In the mid-eighties," says Pat Hurley, "the BuRec and weather modification became rather an unpopular subject for the federal government. They ceased any sort of research. Just cut it cold. We had to do something different."

At about that time, the FAA was developing what it called NEXRAD, for next generation radar. It encompassed a digital radar to allow Air Traffic Controllers to recognize weather features—such as microbursts or tornadoes—as well as airplanes on their monitor screens. It was to operate through computer based algorithms.

Sensing an opportunity, Odegard began chatting up his FAA and Congressional contacts. Odegard made a proposal to research the question of whether or not NEXRAD could detect aircraft icing, a well-known problem in North Dakota and elsewhere. The oddball idea had come from Pat Hurley, still working for the BuRec and still a North Dakota homer. He thought the same process used for cloud seeding could be applied to icing research.

"In cloud seeding, you're trying to identify areas of super-cooled liquid water in clouds that ice-up silver iodide particles," says Hurley. "That's very similar to icing up an aircraft wing. The ideal conditions for cloud seeding are the ideal conditions for aircraft icing. And here we had all this experience studying cloud seeding. Why not apply it to icing? It's an oversimplification, but John understood it. And he sold it to the FAA, with some political pressure."

As a result, about $3 million in NEXRAD funding headed for North Dakota. Tony Grainger, then chair of the new atmospheric sciences department, got the assignment to study whether NEXRAD could identify severe aircraft icing. He wasn't happy about it because, unlike Hurley, he thought it was a crazy idea.

"Somebody sold this NEXRAD thing as being able to tell where there was icing in the clouds," he says. "I don't know who thought they could do that. You're looking for super cooled liquid water, and you can't see it on radar. We were supposed to come up with algorithms to do this. And we did have an algorithm, but it had a high false alarm rate. It was not implemented as part of NEXRAD program, but we worked on it for long time."

Hurley is not apologetic for his theory. "I've been criticized for it," he says. "Scientists say it's over simplification. But it started the whole ball rolling to get involved with NEXRAD—a multi-billion dollar program."

Which, of course, was the whole purpose—get that foot in the door and pretty soon you're in the kitchen. In this case, Hurley estimates that the foot in that door led to $34 million in various FAA funding over the next two decades—not bad for proving a crazy idea wouldn't work. Or for sharpening Grainger's image of an Eskimo buying a freezer from John Odegard.

In any event, proving a negative—no, it turned out, NEXRAD *couldn't* detect icing conditions—came with little glory. So Odegard got into a more interesting and, as it turned out, useful aspect of NEXRAD research. It involved the detection of microbursts—those intensely powerful downbursts that can materialize in seconds. In the mid-eighties, microbursts had caused a number of serious airplane accidents. Could NEXRAD detect them in time to warn pilots away from them?

Grainger's Citation crew worked closely with Lincoln Labs of M.I.T., which had developed algorithms to be tested. Both were using Doppler radars in their work. In August of 1985, not long after they began their research, a Delta Airlines L-1011 crashed in Dallas with heavy loss of life. It had been caught in a severe microburst. The pressure was now on the researchers. Ten days after that accident, Roger Tilbury, the pilot of the UND Citation, was able to fly the research jet directly into a microburst, with all instruments active and the Doppler radar recording everything.

No one had ever done that before (intentionally and successfully), and the FAA was ecstatic—not so much that Tilbury, his plane and crew survived, but because there now existed definitive data to analyze.

The research then shifted to capturing microbursts as they actually formed. But for those two Doppler radars—UND's and MIT's—and the algorithms to work properly together, they needed to be looking at the same cloud from opposite directions, 90 degrees apart. One of Pat Hurley's first jobs, when Odegard finally hired him away from the BuRec, was to scout radar sites near the Orlando International Airport and near Denver's old Stapleton airport and then see if they could spot microbursts developing.

The experiment worked in both places, though in Denver, the researchers got even luckier. "One day," says Hurley, "we picked up a microburst on our radar. Within seconds, we alerted the MIT radar. They verified it on their computer and flashed it to the Denver control tower. The microburst was right over the runway at the same time a DC-10 was coming in. The pilot veered off right away. If he hadn't, he would have crashed. That pilot was so thankful. He said, 'This really works. By God, it saved my aircraft.'"

Once they had proved the system worked, the FAA lost no time setting the radar up at 47 major airports in the country, automatically providing warnings and alerts to pilots. "There hasn't been a wind shear aviation accident for some years," says Mike Poellot, who succeeded Jeff Stith in 1999 as chair of the department. "They've identified a hazard and put the equipment on the planes. We were part of that."

The quality of the box matters little.
Success depends upon the
man who sits in it.
Manfred von Richthofen

Chapter 18
An Historic Day for Luggage

To the casual observer, the year 1986 looked like a pretty good one at the brand new Center for Aerospace Sciences. A nationally renowned astronaut had joined the faculty, the UND Flying Team had won its second national title, the school had extended its reach to Morocco and excited high school students from around the country were attending summer aerospace camps in Grand Forks. And even though veteran George Hammond had retired in 1984, Odegard had managed to lure Bill Shea away from FAA headquarters to chair his aviation department. Don Smith had been moved up to associate dean, enrollment was rising and the wild blue yonder beckoned.

Such milestones may have impressed the landlubber, but for John Odegard, a man who always wanted more and wanted it fast, and who didn't take kindly to the word no, 1986 was actually a very rough year. If it hadn't been for a Canadian-born college dropout named Bob Buley, 1986 would have been no fun at all.

To begin with, it was the year of the last straw for Cessna Aircraft, which more or less threw in the towel on building single engine, piston-driven airplanes. For years, Cessna not only had been UND's main source of aircraft and the major supplier of single engine planes to other college aviation programs, but it produced one out of every two new general aviation airplanes in the world. In 1981, of 12,000 aircraft built world wide, Cessna manufactured 6,000 of them. About half were the high wing 152 and 172 models so popular in the learn-to-fly market. The company had set up more than 1,000 pilot training centers around the world with its own curriculum—a training system John Odegard had adopted in the early seventies.

In the early eighties, once his program started bringing in outside funds, Odegard gave up his patchwork system of leasing planes from doctors, lawyers and friends and threw all of his fleet business directly to Cessna. Certainly, it was much easier and cheaper from a maintenance perspective to deal with a single manufacturer.

Odegard developed a close relationship with Cessna's chairman, Russ Meyer, who had developed a leveraged-lease program to make it easier for colleges to build their fleet. Cessna would sell a plane to a finance company, which took the depreciation—it did a non-profit like UND no good anyway—and then passed the plane along to Odegard at a lower cost. With orders of 30 planes at a time, UND quickly became Cessna's second largest collegiate customer, behind Embry-Riddle.

But in 1981—Cessna's most profitable year with a force of 14,000 on the manufacturing line—the nation reeled under a recession. Oil prices went up, interest rates flirted with 20 per cent and businesses fell hard. Four years later, Cessna produced only 500 single engine planes, with only 2,000 being produced worldwide. Consumer interest in the pricey learn-to-fly hobby fell from a quarter million enthusiasts to less than 100,000. People with disposable income started buying motorcycles instead of airplanes.

Even so, the economy and the price of fuel wasn't the biggest problem.

"The real killer," says Cessna's Meyer, "was product liability." Frivolous lawsuits, he says, cost the company so much money to defend that it became impractical to keep going. Meyer cited a particularly galling case: Two men were flying a 1947 Cessna 140, but had forgotten to check their fuel. They ran out of gas and crashed. Neither was injured, but they sued Cessna, claiming negligence.

"The lawyers got greedy," says Meyer. "There were two or three dozen world-class ambulance chasers who would sue anybody for anything. Anytime there was any kind of an accident, they'd file a lawsuit. Whether you had to pay or not, you still had to spend money to defend yourself."

By 1986, Cessna had quit making piston-driven airplanes altogether, deciding to put all its efforts into manufacturing their Citation jets. Other general aviation manufacturers found themselves in the same boat, to mix a metaphor. Piper wasn't building piston-drives—although it would start up again in 1988. Beechcraft was down to making two piston aircraft models but would soon specialize in turbo props for the corporate market. Sales of Rockwell's Aero Commander fell off and others companies such as Grumman—producers of the piston-drive Tiger and Cheetah— and Mooney, were sold.

"There really wasn't a lot going on in general aviation," says Kent Lovelace. "By 1986, it was dead, dead, dead. There were thousands of planes out there, but no new planes being sold."

It would be like that for most of the next nine years, until Congress passed a tort reform bill in 1994 making it harder to sue an airplane manufacturer. Still, in 1986, the problem "really left John in the lurch," says Roger Martin, an early Odegard graduate and today a Cessna executive. "From John's perspective, the right training airplane was a Cessna high wing airplane. UND put so many hours on their planes that when we discontinued production, John knew the Cessnas he already had would only last a short period. He actually came to Wichita and met with Cessna and pleaded his case: Please build me X number of Cessnas. It just didn't make business sense to do that. It just killed him."

A worse wound would soon follow.

With his second building already dedicated and open, Odegard in 1985 set his sights on a third structure, this one very unlike the other two. The FAA had announced it would move its longstanding management training school out of Oklahoma City and was looking for a new site. With the success of his program and with all of his contacts, Odegard was sure he could win the bid. To assist in the preparation of the bid, Tom Clifford put in a call to Kent Alm, an old friend of his and Odegard's.

A North Dakota native, Alm had earned his undergraduate degree, and later his Ph.D. in education, at UND. In 1966, while a student there, he decided he wanted to learn to fly. The man who gave him his first lesson was John Odegard. Alm went on to the faculty at the Minnesota State University at Mankato where he served as a vice president and interim president. While there, he moonlighted as a Learjet pilot for Carl Pohlad, the owner of the Minnesota Twins.

Alm returned to North Dakota and was appointed by the governor to the Board of Higher Education. The board then named him the state's Chancellor of Higher Education. In that capacity he had plenty of contact with Tom Clifford who was a board favorite. After his term, Alm became an education consultant in Bismarck, and it was there Clifford reached him in 1985 and asked him to help put together the FAA bid proposal.

"We put together the best proposal the FAA has ever seen," says Alm, now retired. "We were sure we were going to get it."

Odegard was so confident that he talked his architect friends Bill Schoen and Jim Kobetsky into drawing up plans on speculation. They estimate it cost them $40,000 of their time to design a $10 million, 10-story building complex. "John was such an influence on us we were willing to take a gamble," says Schoen. It was only a small part of the $220,000 Odegard himself spent on researching and preparing his bid, according to the *Grand Forks Herald*.

The FAA asked for bids in the summer of 1985, but delayed its final decision again and again. Congressional delegates from multiple states were jockeying for the contract. At one point, when the decision of who would get the bid was again postponed, Odegard complained to *Herald* reporter Marian Young, "If we don't get it, it will be strictly because of politics." Given the mechanics of his past funding victories, it was a statement hard to argue with.

And hard to take when, on the Ides of March, 1986, the contract was awarded not to UND but to Odegard's personal rival, Embry-Riddle University. President Ronald Reagan made the decision, purely political, to support a Republican in Florida running for re-election to Congress.

The news broke on a Friday afternoon. Diane Odegard remembers John calling her at home from his office. "We had plans to go to some function," says Diane. "But he called and said, 'We lost this thing. It's going to be on the news. I'll be here 'til late. We're not going to such and such.' That was uncommon. We pretty much went to every event. It was a huge, huge disappointment for him. He wanted it so badly. He was very down, very sad, very shaken by that."

The third leg of the triple whammy served up that year came in November of 1986. Mark Andrews, the man who had beaten the pork barrel for so many years on behalf of North Dakota, was defeated in his bid for re-election to the U.S. Senate. He lost to Kent Conrad by 1,500 votes. Although Conrad would later prove a solid supporter of aerospace, he at the moment didn't have the immediate standing in the Senate as Andrews.

Friends say this was one of the few times they'd seen Odegard less than his enthusiastic self. Yet as usual, he flexed his ability to pick himself up from a dead stop and move on to the next project. For among the wreckage of so many hopes, there was already a doozie in the works that would more than make up for it.

One of the roots of that doozie goes back to the days in the late seventies when a young Navy veteran named Bob Buley was a pilot for North Central Airlines, headquartered in Minneapolis. A native of Canada, Buley had attended the University of Wisconsin in Madison after mustering out of the U.S. Navy. He then left college because he wanted to learn to fly airliners—a goal, he was warned, that was stupid. But he made it anyway and was hired by North Central. When he wasn't flying, he instructed and worked on technical projects for the company.

In 1979, North Central and Southern Airways merged to become Republic Airlines. One day, Buley's boss asked him if he'd give a tour to a group of visiting aviation students from the University of North Dakota. They had just come from a routine 15-minute tour of Republic's rival, Northwest Airlines, whose headquarters was just down the road in Minneapolis. Buley told the students he could do better than 15 minutes and proceeded to pull out all the stops. The students loved his enthusiasm and style. After that, whenever the North Dakota kids came to town for a tour, they stopped first at Republic and asked for Buley.

One year, those students asked their department chair, John Odegard, if they could invite Bob Buley to give the address at their annual family weekend banquet. Odegard guarded those guest speaker slots jealously, always angling for a big shot who could do the school some public relations good. People like Buzz Aldrin, or Russ Meyer.

Bob Buley was a Captain with North Central Airlines in the late seventies when he met John Odegard. They became fast friends. When North Central was absorbed by Northwest Airlines, Buley helped Odegard plan and write the Spectrum program, designed to produce competent airline pilots from untrained students who had never before flown.

Odegard didn't like to be called a "wheeler dealer" but there's little doubt that he thrived on the daily juggling of high-concept projects, and using his charisma to convince others to invest their time and money with him.

"Bob Buley?" said Odegard. "Who the hell is he?"

But the students prevailed, and Buley flew out to Grand Forks to finally see the school he had heard much about. After his speech that night—which many still remember for its high-toned literary references and overall eloquence—Buley and Odegard started chatting and hit it off.

After that, Buley was invited back on a regular basis to teach classes on airline operations. He and Odegard chatted by phone almost every day. "I was bowled over by his enthusiasm," says Buley. "Some people had a negative reaction when they first met John. They think he's a flim-flam man. He was too much to believe. Too good to be true. 'He's going to pick my pocket. That can't be real.' I didn't have that reaction. My reaction was: 'Okay, now I'm starting to put it together.' Because I'd already seen and flown with his students. I knew what the university was capable of. John had such an eye for detecting things in young people."

One of the ideas that went back and forth over the wires from Grand Forks to Minneapolis had been a favorite topic of Odegard's for years. The airlines, he would complain, weren't involved in the education process—to their detriment.

"We both believed," says Buley, "that if an airline could identify kids in their freshman year of college, they'd be able to watch them, get involved in the education process, be able to get them into productive positions in the airline sooner. You'd actually end up with a higher quality product. We thought getting an airline involved to that extent was a worthwhile goal."

Odegard and Buley also saw on the horizon a looming pilot shortage. Vietnam-era pilots who had gone from fighters to airliners were beginning to age. Meanwhile, the military trained fewer new pilots during peacetime, and at the same time, airlines were expanding and in need of pilots.

Buley told Odegard about plans at Republic Airlines to buy a simulator for their new Boeing 757. Those plans had been put on hold because of the cost. The idea the two cooked up was for UND to buy the simulator, house it in Grand Forks and lease it back to Republic. The airline would man the simulator, maintain it and train its pilots on it in Grand Forks. In off hours, UND students could train with it. Odegard saw the simulator being built into an advanced flying course for what was known then in the aviation world as *"ab initio"* training. The Latin phrase means "from the beginning," and in aviation terms, a student would be trained from the point of knowing absolutely nothing about an airplane to being certified as a multi-engine airline pilot. Airlines in Europe and Asia—Lufthansa being the most prominent example—have used an *"ab initio"* approach to "grow" their own pilots for sometime.

"The idea," says Buley, "was that ultimately we'll get airlines to look at these students in the early stage of their education and start evaluating them for potential."

Buley wrote a draft of a letter to his employer, Republic, proposing just such a plan. He sent the letter to Odegard who read it, liked, and signed it. He then mailed it to Republic. The letter, as Odegardian luck demanded, ended up in the hands of Bob Buley's boss, Republic's vice president of operations.

"My boss hands it to me," says Buley, "and he says, 'Look into this and draft a response.' I called Odie and said, 'You're not going to believe this. I got the damn letter in my hands. My boss just told me I have to look into this and draft a response.' John said 'Ohhh Bobby, that's great! Fantastic!'"

A small speed bump: three weeks later Northwest announced it was buying Republic Airlines. Buley kept his job but didn't think it wise for an ex-Republic man to be proposing so soon to his new employer such a grand program like *ab initio* training. Instead, Odegard sent a reworked copy of the same letter to Joe Lapensky, the former president, then chairman of Northwest.

"And Joe thought it was a good idea," says Buley.

Not surprising, since Lapensky had been a fan of the University of North Dakota ever since his old boss, Don Nyrop, had introduced him to Tom Clifford back in the seventies.

Clifford and Lapensky were contemporaries. Lapensky joined Northwest's finance department in 1945 in an old shaving cream factory in Minneapolis. In the sixties, he helped start Northwest's economic planning department. He was vice president for finance when he finally met Clifford. Through Nyrop—and as a favor to Clifford—Lapensky agreed to come to Grand Forks weekly to teach classes in airlines economics and management. He succeeded Nyrop as president of the airline in 1976 and became chairman of the board in 1985.

"John was very much the opposite of Joe," says Bob Muhs, a man who knew them both. "Joe was an accountant in the traditional sense. Very quiet, very reserved. I think he liked the personality, the drive, the eagerness of John. I think Joe thought John was a character."

But Joe wasn't a pushover.

"John was very enterprising, very forward thinking," recalls Lapensky. "And he was very ambitious. I can remember flying around North Dakota with him. He was always asking for more. He wanted Northwest to do a lot of things and some weren't very practical. But he was very convincing. You had to ride herd on him, but he was a great guy, and we got along fine."

When Odegard brought up the idea of a cooperative training program, putting Northwest simulators in Grand Forks and developing an *ab initio* program, Lapensky was intrigued. Like Odegard, he saw a pilot shortage in the near future. He also knew Northwest was expanding and would need fairly soon to train a large number of pilots.

Lapensky was of the same mind as Don Nyrop before him who felt that the training pilots got in the military was insufficient for airline flying. Nyrop believed in university training, says Lapensky, and so did he. He liked that Odegard "didn't just teach flying, but aviation administration, accounting and a kind of business-oriented aviation. Which was a good idea. He turned out a lot of people who didn't become pilots but who were well trained and had a background in aviation management."

So he was predisposed to listen while Odegard spun the dream as he saw it—and as his architects had drawn it up, once again, on speculation. That design borrowed some from the rejected FAA management center, but went far beyond. Odegard described a $200 million complex on the CAS campus that would serve as Northwest's main training center. It would include enough room to house dozens of large, full-motion simulators, multiple classrooms and a 300-room hotel tower. Every bit of it would be wired for computer connectivity—a term that barely existed in 1985—and designed according to the latest thoughts on human factor compatibility.

"It would be the consummate school house," says Buley, who cheered from the sidelines. Until then, he adds, Northwest had no plan at all for building any training facility, in spite of their need. They had already completed a modest training structure in Minneapolis, and the recently purchased Republic Airlines came with training facilities as well.

"It was against the corporate culture of Northwest to build anything that could be classified as extravagant," says Buley. "Their attitude was, 'We'll get by.'"

Instead, at Lapensky's urging, Northwest created an independent company called NATCO—for Northwest Aerospace Training Corporation—to explore new training strategies, including Odegard's Grand Forks campus plan.

In the meantime, the Aerospace Foundation took an option on a 20-acre tract of tax delinquent land across Forty-second Street in Grand Forks from CAS I and CAS II. Clifford and Earl Strinden also convinced the Burlington Northern Railroad to sell the foundation a parcel of land adjacent to those acres. The foundation had been trying to find a developer to turn that plot into a technology park complete with a hotel, but were having trouble enticing top name companies to locate there. In March of 1986, when Northwest announced it would indeed build the training center in Grand Forks, the problems with the technology park seemed a moot point.

Yet four months later, to Odegard's great dismay, Northwest had second thoughts. They would not build their main training complex in Grand Forks, but would hire Odegard's architects to put up almost the exact same complex he had described—minus the 300-room hotel tower—in Minneapolis.

The reason: It would cost too much to ship its pilots to Grand Forks for training, said Northwest. Their labor contract would require pilots to be paid a per diem charge during the training. Suddenly people at NWA began asking, "And just why are we building this in Grand Forks?"

Recently, Lapensky thought back to that time and said, "John's problem was he was in a place that didn't warrant all the things he wanted. It was kind of remote. It was a hard place to get students to come to. I'm surprised it was as popular as it was. His geography was kind of against him."

But Odegard was not an easy man to be *geographied* out of a building. He pressed NATCO and Lapensky for some consideration.

"He was a go-getter," says Lapensky. "A gadfly."

So much so that four months after it said "no," Northwest consented, almost as a consolation prize, to partner with the FAA and share the $6 million cost of a new CAS building. It would be a training center of sorts, a branch of that new NATCO training center under construction in Minnesota. Northwest planned to install a DC-9 and 727 simulator in its half of the new CAS building in Grand Forks and assign several staffers to help with student training and development. The other half of the building was to be used for CAS classrooms and faculty offices.

"It's time for Grand Forks to reassess its world position," said a rare boastful editorial in the *Grand Forks Herald*. "We are accustomed to thinking of our area as remote, but the facilities at UND put us at the gate of the aviation industry and of an important part of the future."

Odegard, however, wasn't waiting for the future. He wasted no time waiting for another last-minute second thought by the airline. He promptly staged a ground-breaking ceremony for the new building, even though the final details of Northwest's involvement had yet to be worked out. It was a curious ceremony, with Earl Strinden on hand that day in September of 1987 to offer remarks to the assembled dignitaries. They included many Northwest executives.

"This is an historic day in Grand Forks," Strinden told the assembly. "A rare day, in fact. Because the luggage for the Northwest executives here today arrived on the same plane they did."

If some people tittered, it wasn't the gang from the Gopher State. Years later, though, Lapensky would chuckle that Odegard had pushed so hard and so unceasingly to get his third building that it caused grumbling in Minneapolis. Some complained that the meaning behind the acronym NATCO had been permanently altered. From now on it would stand for *Not Able To Contain Odegard*.

It was a joke so funny that Odegard's campus critics had already forgotten to laugh. Even though they were quietly pleased when Odegard had lost his bid for the FAA training center, they groaned noisily when he won something that was sure to make him even more insufferable than ever. For Odegard had asked Kent Alm to stay on after the FAA defeat and become his director of program development. His first task was no minor matter: get the Center for Aerospace Sciences the authority to grant its own degrees.

Alm got to work and did such a convincing job in the writing of his proposal that Tom Clifford simply bypassed the normal procedure of going through the

Faculty Senate. He sent a request directly to the state Board of Higher Education to change CAS from a center to a degree-granting school. When it was approved, as he knew it would be, Clifford then anointed Odegard, heretic of all the straight and narrow, as a dean of the realm.

The two prime movers in the universe
are time and luck.
Kurt Vonnegut, Jr.
Hocus Pocus

Chapter 19
White, Clean, Fresh, Impressive

Even though Northwest had decided against Grand Forks for its main training center, Joe Lapensky still liked Odegard's idea of an *ab initio* program. Northwest knew it needed to become proactive in pilot development rather than rely on a shrinking supply of experienced pilot candidates.

"John sold Joe the *ab initio* program as an investment in the future," says Bob Buley. "We were pretty high during that time. We'd get on the phone almost everyday asking each other, 'How's it going? What happened today?' I remember John saying, 'Oh man, this is so good, so awesome.'"

Northwest and the Aerospace Foundation jointly funded the research for the *ab initio* program; the airline even made several of its line captains available to work on the curriculum. They worked through Kent Alm, promoted to head of program development, to plan and to write a start-to-finish curriculum that would turn a complete aerodynamic rube into a competent airline pilot. Twenty people gathered at CAS in late 1986 for the first meeting of the *ab initio* group. There, the project was officially dubbed Spectrum. A core group did most of the work over months of four-day-a-week meetings in Grand Forks and Minneapolis. They included Buley, Kel Moeller and Bob Patzkey—all Northwest captains; Paul Rapicky, a software expert from the Wyoming Centrum for Assistive Technology (WYCAT); a pair of human factors experts—Dick Jensen from the Ohio State University, and Stan Trollip; and the usual suspects, Al Palmer, Dana Siewert, Don Dubuque and Kent Lovelace of the CAS department of aviation.

Northwest had insisted that the group completely invent its own program and not borrow bits and pieces from existing curricula. It meant that the basic philosophies and mechanics of flight and flight education had to be re-examined, agreed on and charted one by one.

"We started on day one," says Al Palmer. "Okay, a student pilot walks in the door. What does that person need to know? It was a year's worth of just contemplating your navel, but it had to be done. We had the knowledge. This was a new program and it was going to be taught in a fairly new and innovative way."

Spectrum differed from the regular pilot training taught to Odegard's aviation students.

"It was a major difference in curriculum," says Alm. "It was the world's first *ab initio* program for pilots with an airline mission, which meant they would be trained in pairs. All other pilots had single pilot training. But these pilots were going to go into commercial airliners as part of a crew. So we started in the first hour in the simulator with a pilot and a copilot."

"It's one thing to be able to fly an airplane," adds Buley, "but another to have the skills to be able to work as a crew member."

There were some rough spots. For one thing, Bob Buley's status as an ex-Republic captain made him something of a second class citizen in the eyes of the veteran Northwest pilots. "Northwest pilots just weren't too anxious to admit that Republic could even fly," says Kent Alm. "But Buley was a bright, outgoing guy and a good pilot."

One of the other Northwest pilots, "a crusty old captain," says Buley, was a heavy smoker. Though Odegard had always banned smoking in his offices, this captain continued to smoke anyway. "It just drove John nuts," says Buley. "But this guy would say, 'You want me to do this?'"

At other times, when energy flagged, Odegard's cheerleading would reinvigorate the hard working group. "He would infect you with his enthusiasm," says Buley. "I don't think it was contrived. It was just the way John dealt with things."

Twelve months after they started, the first phase of the mission—a job task analysis—was complete. It consisted of 10,000 lines of text in each of four books. Each manual detailed a type of airplane. These books became the foundation of the Spectrum curriculum which the group could now start writing.

In the meantime, while Spectrum was viewed by Northwest as a potential tool of the future, Odegard was working feverishly behind the scenes to use it to bring in a major international client right away. That odyssey started when one of Odegard's graduates in the early eighties landed a job in Taipei, Taiwan, with China Airlines. His boss was one of the loyal old Nationalist officers who had fought under Chiang Kai-Shek during World War II and in the civil war afterward, retreating ultimately to the island of Taiwan. Retired general Shan Che-Tao was then a vice president of China Airlines, and he too faced a pilot shortage in the face of expansion plans.

At first, General Shan wanted to train eleven of his airline's flight engineers as pilots. He'd been looking around for a suitable program when the UND alum suggested he investigate North Dakota. He did just that. Contacts were made in

When retired general Shan, Che-Tao, a vice-president of China Airlines, faced a pilot shortage in the face of expansion plans he turned to North Dakota and John Odegard's aerospace program to train new pilots.

Probably the greatest coup of John Odegard's career was landing a contract with China Airlines to train raw students as airline pilots. The lucrative contract, executed by the independent Aerospace Foundation, built an international reputation for Odegard and his school.

1984—as Odegard was being courted by Embry-Riddle, and the Morocco project was just coming into view. Odegard and Don Smith met that year with China Airlines officials in California and in Taiwan. By 1985, the prospect of China Airlines as a client looked promising enough to remove any doubts about the creation of an Aerospace Foundation.

After initial talks with UND, General Shan altered his plan to transforming 12 flight engineers and 12 flight attendants into pilots. Negotiations went on for a couple of years under the auspices of the foundation, which, by now, had gotten its feet wet with the Moroccan adventure. Odegard was now flying as high as he ever had. Still, he never stopped reminding his staff that the devil lay in the details, and that without close attention to all of the little things, the China Airlines deal, the Spectrum program, the Northwest connection, could collapse.

"A very big part of John's success was making an outstanding impression on people who came to North Dakota, not knowing what to expect," says Mike Poellot. "They would see these facilities, and they just couldn't believe it. Visitors were just continually amazed. So our work areas always had to be cleaned and spotless. John wanted everything to be looking good for whoever he brought by."

Al Palmer saw that from the day he started in 1976.

"We had 12 airplanes, but when we took a picture of our fleet we'd have those things lined up, and we made them look big. We always did things in a grand way, like a cat that gets scared and fluffs her tail up to make her look bigger. If we had a trailer, that trailer looked great. John never settled for second best. When we ordered airplanes, we ordered the best planes we could for that time. We always had state-of-the-art equipment."

With aviation people, it was fairly easy for Odegard to get his message across. But when it came to scientists, Odegard didn't always get what he wanted. He once tried to fire a scientist because he felt the man came across as meek and withdrawn.

"This man was a very productive scientist," says Tony Grainger. "But he was not a salesman. He didn't toot his own horn, didn't put forth a great presence. John concluded this guy didn't seem to be doing much, so he didn't renew his contract." Grainger managed to change Odegard's mind when he insisted the man was doing work no one else could do.

Grainger recalls another butting of the heads where the scientists emerged with a morale victory. He and his team were working in a hangar with one of the aircraft. "We didn't have the right kind of power unit for the plane," he says. "We needed 28 volts. We thought, 'Well, what could we do?' We went out and got some truck batteries and a battery charger and a little cart and a cable. We put it together and plugged it in, and it worked fine."

But Odegard spotted it on one of his inspection tours and was dismayed.

"John looked at it and said, 'Oh man, that's really ugly,'" says Grainger. "But to a scientist, that looked beautiful; it did the job for a minimal cost. To John it didn't look high tech. He wanted us to buy one of those DC power carts. He was trying to spend $5,000 to do the same thing. We said, 'John, if we do that, everyone will say, "Man you guys have money to burn."' But this junky collection of truck batteries didn't give off the right impression for John. He cringed whenever he looked at it."

And look at those facilities he did. No matter how busy, Odegard would always find time to walk through the hangars and maintenance bays for a quick inspection. Like a drill sergeant, he could always find the one spot of dust someone had missed.

Frank Argenziano, the head of maintenance at the airport, was in tune with Odegard's insistence on looking good. In fact, a choice he'd made years earlier played a significant role in landing the China Airlines business.

"Something I always hated," says Argenziano, "was a messy shop. It wasn't professional, and it didn't make for a good operating environment for a mechanic. When I came here, guys were scruffy looking. The shop was dirty. When we built the new hangar, it had a yellow grungy floor. I got plant services to put down a coat of white paint on the floor. I was amazed at how bright the place got. You could see if stuff was laying on the floor—nuts, bolts or dirt. When we did the next hangar, we just put a clear coat over concrete. It was dark. Then we decided to build another hangar, and I said, 'Hey, lets put some money into this and get it painted.' John said okay. So we ended up stripping the floor down, and a company came in and painted it white and put a coating of polyurethane on top. With those sodium vapor lights shining down, it looked like a showcase. It really made a difference in the attitude of everyone there."

In January of 1987, it made another difference. The negotiations with China Airlines had reached the point where the president of the airline, General Wu Yueh Chan—another of Chiang's old Nationalist deputies—decided to come to North Dakota to see if things really were as Odegard had described them.

On the day he arrived, it appeared that Odegard's famous luck had gone south. A ferocious blizzard had hit Grand Forks, covering everything with massive drifts of blowing snow. Odegard was flying the general to North Dakota aboard the UND Citation. But there was so much snow that once the jet landed at Grand Forks Airport, it got stuck on the taxiway. There was nothing to do but disembark the passengers and trudge through the snow to the first building in sight. It happened to be UND's Northeast maintenance hangar. Odegard groaned when he realized all of his carefully laid plans for greeting General Wu in a warm, neat reception area, perfectly appointed with first impressions in mind, were in the wind.

"Oh my God," he mumbled. "Here's the chair of the entire airlines. He's never going to come to North Dakota."

But when General Wu stepped into the hangar his eyes brightened as he took in the white floors.

"Ooh," he said. "Just like China Airlines. White, clean, fresh, impressive." He then turned to Odegard and said, "This weather is just like northern China. I feel like home. This is my home."

"We couldn't have planned it any better than that," says Palmer. "And as a result of that, we got a contract that went on for 10 years. I hate to think how many millions of dollars those white floors brought into our program. The contract made us international. And *that* was the big moment."

Impressed with what it found in North Dakota, China Airlines began discussions about using Odegard's facilities to train not just two dozen but hundreds of pilots over a period of years not months. It meant that the volume of air traffic and students in Grand Forks was about to increase, taxing all existing facilities and aircraft. Even General Wu saw it. Though convinced he was dealing with the right people, he made clear to Odegard that to make the contract work, the UND airport facilities—clean as they were—would need a significant upgrade.

Don Smith came up with a list of necessary airport improvements, including new hangars, a new maintenance facility, the purchase of the buildings that were being leased, remodeling existing structures and construction of new tower. Bob Reis figured the cost at $17 to $20 million, including an upgrade of the aging Cessna fleet. Reis and Irv Walen spent long hours putting together a five-year business plan aimed at drawing investors.

It was Reis's idea to raise the millions by floating tax exempt municipal bonds to private investors. The plan was announced in March of 1987, and by the end of August, $17 million had been committed to the bonds. Further, the city of Grand Forks agreed to undertake the remodeling project through its newly established airport authority. Construction began even as Odegard and Don Smith were flying back and forth to Taiwan, hammering out the specific details of the yet-to-be-finalized China Airlines contract.

In the middle of all of that, Odegard, and a retinue of pilots, accountants and university administrators flew down to Vero Beach, Florida in 1987, to the headquarters of the Piper Aircraft company. With Cessna out of the single-engine business, Odegard needed a new supplier of aircraft. In fact, he was looking for someone who could sell him at least 150 airplanes. Enrollment at the school was approaching 2,000 students, China Airlines was sending hundreds more, and Odegard was hatching plans in another corner of his busy mind to open new campuses around the country.

On the flight down to Florida, however, he casually told Lyle Beiswenger—the VP for finance who had done some of the price negotiating with Piper—that he'd decided to buy an extra 100 planes for a total of 250. By now, used to financial surprises from Odegard, Beiswenger simply rolled his eyes. "Well, John," he said. "Whatever."

By the late 80s the FAA had built a new tower adjacent to
the UND airport facility. The old tower was relocated not far
from the old Quonset hut barn (upper left) and was used by
Odegard's air traffic control students. By then, several new
hangars had been built and the barn had been refurbished
and connected to the rest of the facility.

The first class of China Airlines students arrived in North Dakota in the fall of 1988. The
Spectrum curriculum designed to train them was barely off the printer when classes began.
The program eventually expanded from training in piston drive air craft to jet airplanes and
greatly expanded the UND fleet.

Piper, at the time, had changed ownership and was trying to revitalize the company by re-starting production of single-engine planes. At the time, the company was producing some twin engine propeller craft and even some twin engine turboprops. Its bread and butter, however, was the low-wing piston-engine plane such as the Cadet and the Arrow. It was Odegard's only practical alternative to Cessna.

Piper executives were keen enough on getting the UND business that they agreed, according to Buley, to make minor design adjustments in his single engine models. Piper had introduced a T-tail design on the planes. But after Buley, Don Dubuque, Al Palmer and Dana Siewert test flew several models, they convinced Piper to go back to the older configuration.

At that point, Odegard and Piper agreed to a potential $29 million deal that gave UND the sweet option to buy 250 Seminoles, Arrows and Cadets over the next five years *at the same fixed price*. That deal, which had been negotiated ahead of time by Don Dubuque, included a guaranteed price for all warranty maintenance costs over the contract period.

Odegard would note when he returned to Grand Forks that this was the largest purchase of aircraft in the history of general aviation. But even though he and Piper had agreed to the deal, Odegard was clever enough not to actually sign it. "John had been making all the decisions," recalls Beiswenger. "But when it got down to putting his name on the line it was, 'Oh Lyle, you're the VP. You better sign it, not me.' I got kind of a kick out of that."

A footnote: While the deal turned out to be a good one for UND, it may have been the deciding factor that put Piper into bankruptcy proceedings two years later. In the fall of 1990, Piper laid off half of its manufacturing force. It had supplied UND with 55 planes and owed it another 26 according to that stage of the contract. Piper would eventually right itself. It continues today to be the principal supplier of single engine and twin engine aircraft to the UND program.

Back in Taiwan, in the spring of 1988, Don Smith was carrying the brunt of the negotiations with China Airlines. He had quite a selling job to do. The Chinese had looked at the *ab initio* training model of Lufthansa Airlines, which charged about $40,000 per student. Smith wanted $84,000 per student.

"The Chinese were astounded," Smith recalls. But he walked them through a very detailed training plan, noting all of the elements the Spectrum group back in Grand Forks were working on: psychological testing, cockpit management, competency in the English language, radio protocols etc.

"I told them this would be your program. Not our program. We will provide the curriculum and the individuals to make it happen. But we want you to send one of your captains to oversee it, to make sure you're happy with what we're doing."

Smith then had Dana Siewert write onto a blackboard each element in the proposed UND program. "It was kind of a strategy to sell them," says Smith. "I said to them, 'You tell me which of those elements you want to eliminate.' Well, nobody wanted to give up anything."

In Grand Forks, in late summer of 1988, Buley and Palmer and the others were finishing up that year-long task analysis in preparation for writing the actual Spectrum curriculum. They thought they had another six months to finish. But Odegard appeared in their midst one day in August of 1988 and dropped a bomb. The deal was done, and the first class of China Airlines students would arrive in Grand Forks in just six weeks.

"We were all like, 'Holy shit! John what are you doing?'" says Buley. "But John had told China Airlines it would be done. Which meant it would be done. If he worried about it at all, he was fairly close mouthed."

The group barely had time to worry as they moved immediately into warp speed. "There were a lot of long days and a lot of long hours," says Dana Siewert. "I don't think anyone resented that. We took it on as a challenge."

Those first 24 trainees from China Airlines were far from the average student. For one thing, they already had college degrees. "The task," says Siewert, "was how to build a program and give training when they didn't need a college education, yet that was what the university was there for. How do we do that? How do we provide the same quality of training that we do with our undergrads, but in 18 months, not four years? We didn't know how, but John had already convinced them that we did. He's wined and dined them, given them the dog and pony show. He's sold it, and they're coming. It isn't that we need to get back to them and let them know if we can do this. It's 'They're coming guys and they'll be here in October.'"

Those six weeks went by all too fast. "We built the first block of the curriculum, about five lessons, and the students came in the door," says Palmer. "As they were taking those first five classes, we were building lessons six to ten. The students would get ready to go on to the next block of training, and we were getting their course outline fresh off the printing press. We were one step ahead."

Inevitably there were glitches in those first outlines, most of them due to unforeseen cultural differences.

"These were smart, intelligent people," says Siewert. "Many had master's degrees in engineering and accounting. But in Taiwan, their primary means of transportation was the bicycle. Many of them not only didn't have a driver's license, they didn't know how to drive cars. We're teaching them to fly complex airplanes in a multi-dimensional environment, and they can't drive a car. That was a pretty big undertaking."

Other problems cropped up, things no one had thought to ask about ahead of time. "We learned that just because they spoke English didn't mean they could fit into an aviation environment," says Siewert, "because there's a difference between aviation English and the typical English that you and I know. When we talk about 'crossing the outer marker at 2300 feet,' or 'don't lower the landing gear until the command bars start,' the typical person on the street wouldn't understand. There were words they had never heard before."

So the Spectrum group devised a course in aviation English, introducing students to terminology everyone else took for granted. "They would hear it in English," says Siewert, "and it took them seconds to convert it to Taiwanese. Then they had to speak it back in English again. There was a time gap. When you're flying 200 feet above the ground doing 100 miles an hour, you just don't have a whole lot of time to start transitioning words. That was probably the biggest challenge."

There was one other small matter to consider. Though the contract was now set, the first $400,000 payment from China Airlines wouldn't be made for three months. While Odegard had committed to starting the program, he actually had no money at all to get it up and running. The new fleet of Piper planes and the new facilities were all scheduled to be paid for by the bond issue. But those soon-to-arrive students needed computers, food, housing and other services. Salaries had to be paid and classrooms furnished.

Don Smith saw the problem right away but says he couldn't get Odegard to focus on it. He made a detailed estimate of what it would cost to start the program. "Realizing we'd made a commitment up front," he says, "I went to the bank on behalf of the foundation—which had no money in the till at all—and I personally borrowed $200,000 as seed money to get the program going. John never worried about how he was going to pay for it. But it always got done."

PART THREE

WINDSWEPT HEIGHTS

Up, up the long, delirious burning blue

I've topped the windswept heights with easy grace

Where never lark, or even eagle flew.

by John Gillespie Magee, Jr.
High Flight

How can life grant us boon of living, compensate
For dull gray ugliness and pregnant hate
Unless we dare?
Amelia Earhart

Chapter 20
West of the Coulee

Everything seemed to come together that October of 1988 when the Chinese students came to Grand Forks. It was exactly twenty years since John Odegard had welcomed his first handful of aviation students with lots of bravado but no guarantees, little support and even less money. In two decades, the handful of students had become 1,800. Ernie Fox's two planes had become 70, plus two jets, four helicopters and 130 instructors. The cramped offices of Gamble Hall were now a formidable complex of three sophisticated space age buildings, with a fourth in the works.

The only thing that hadn't changed was Odegard's bravado. He talked of linking his buildings with elevated, enclosed walkways that he took to calling "gerbil tubes." He even raised the idea of buying the monorail that had been used to transport patrons at the Montreal Expo in 1967. His idea—never realized—was to set up the monorail to shuttle UND students, especially medical students, from the main campus to the recently opened medical park on Columbia Road, south of the university. Of course, it would also extend to the aerospace complex.

A narrow stream known as the English Coulee meanders between the sprawling CAS complex and the rest of the spread out university. Even today, many on the main campus condescendingly describe aerospace people as being "west of the Coulee." Odegard felt the sting of that isolation, and he didn't like it. As he was having his impractical plans drawn up to relocate the monorail to Grand Forks, his close friend Ken Svedjan remembers him defending the idea saying, "We need some connectability out here."

Connectability seems to be a recurring, motivating theme in Odegard's life. He was a man of elaborate connections, from the students he loved, to his network of loyal alums, to the big shots with deep pockets and the small shots who knew the big shots.

During one of his visits to the UND campus, General Wu Yueh Chan, president of China Airlines (second from left), is hosted in a round of golf by Bob Reis (far left), airline Capt. K.C. Liang (second from right), and Tom Kenville.

John Odegard's enthusiasm and charisma was such that often he was described as being larger than life. In this picture, during one of his trips to India, he even seems larger than the Taj Mahal.

"He didn't bank on raising money," says Svedjan. "Instead, he went to the Congress. He knew how to develop the linkages. It didn't matter Republican or Democrat. John was smart enough to know you need to butter your bread on both sides. He was very successful at that."

One linkage Odegard never ignored was the connection to his parents. Odegard remained close to them wherever he went, dating from his brief adventure at the University of Denver in 1959. While there, scrounging for a way to support himself while attending classes, he wrote a touching letter to Clara and Truman. It was a son's surprised and humble recognition that perhaps he'd bitten off more than he could chew.

"I'm more than thankful for the money you've sent me," he wrote in the fall of 1959. "It makes me feel awfully cheap to ask you for money, but I guess that can't be helped. I didn't realize how much I would need you both and how much I depend on both of you until now. I don't know what I could do without you two."

Odegard was always a prolific writer of letters and cards. On his trips away from Grand Forks, he would send touching post cards to his children—notes of praise and encouragement and expressions of love. Diane remembers that wherever he went, and however long he was gone, her husband kept in constant touch with her.

Connectability certainly explains his addiction to networking and his early adapter attitude to any form of communication. He had a car phone before most, had one of the first cellular phones in Grand Forks, and one of the first facsimile machines. His home and his family cabin on Pike Bay in Minnesota had telephones in almost every room. His friends and family remember him always on the phone—most, when asked about it will repeat the word "always" for emphasis. More than once, more than a hundred times, those who had a meeting scheduled with Odegard would come to his office only to find him on the phone. He'd seldom cut the call short. He was more likely to carry on a lengthy conversation, hang up and say to his visitor, "Just one more call." Few can remember when that didn't stretch into two or three calls. It may even explain why he stayed in North Dakota, the bull's-eye of all of his connections. All he had to do was hop on one of "his" planes, and he could connect to any dot anywhere on the globe.

"People couldn't get enough of him," says Jerry Murray. "The UND folks thought he was in demand too much. He was gone a lot, traveling all over the world, building relationships and loving it. He was ten years ahead of globalization. When he'd come back, you'd sit and listen to him tell about everybody he'd met. He loved bringing those people to Grand Forks. And sometimes Grand Forks wasn't quite ready to see the diversity Odegard was bringing to town."

Ironically, for all of his connections to so many people, Odegard once confided to Diane that he felt as if didn't have any close male friends.

"It was one day we were in the car," she recalls. "He said, 'I can't think of one close friend.' Well, he had tons of friends. He was close with other men. But he said he never felt he was close. I think he was saying it with regret. He knew everybody, but he didn't do things with other guys. He didn't run out to play golf with his buddies. People he would confide in? No, he didn't do that. But other people came to him and confided in him."

Even so, things really were coming together that fall of 1988.

The third CAS building was finished and would soon receive its two airline simulators from Northwest Airlines. Construction would begin in December on a fourth building, a five-story tower of classrooms and administrative offices at the airport for the department of aviation.

At Buzz Aldrin's suggestion, Odegard had hired David Webb, a Reagan appointee to the U.S. Space Commission and an expert in hypersonic flight. In 1987, he founded the department of space studies at CAS. A master's level graduate program, the aim of space studies was to apply knowledge gleaned through space exploration to resolving social, political and scientific concerns on earth. Along those lines, Odegard was promoting a fourth CAS building to house the space studies department along with a sophisticated computer operation. Called the Earth Systems Science project it was envisioned as a joint project between aerospace and the department of geography in the School or Arts and Sciences. Odegard's hope was to get the National Aeronautic and Space Administration to fund it to the tune of $20 million—just for starters.

Right away, Webb began setting up an international conference on hypersonic flight in Grand Forks, the first of its kind anywhere. In September of 1988, he brought to town world leaders in advanced engine and airframe design, and the integration of advanced computer flight control systems. Most of these experts had been working toward an ultimate goal of building the world's first hypersonic aircraft.

Meanwhile, Odegard had gone to Quentin Burdick, North Dakota's senior U.S. Senator, for help with the funding of that Earth Systems Science—or ESS—project. Burdick was a member of the Senate Appropriations Committee and had strong ties to the U.S. Department of Agriculture. He'd used those ties to secure a $10 million USDA grant to build a basic science research building for the UND School of Medicine—a school with which he'd had longstanding connections. Yet until Mark Andrews lost his Senate seat in 1986, Burdick had never been west of the Coulee. That year Odegard went a courtin' and sent a plane to Fargo to bring Burdick to Grand Forks. On seeing the aerospace center for the first time, Burdick told a reporter, "Anybody that wouldn't be impressed by this is dumb."

Burdick then approached the USDA about funding the Earth Systems Science Institute—as it was now called—to the tune of $8 million. In September—right around the time Al Palmer and Dana Siewert were scrambling to write their Spectrum curriculum, and scientists from Britain to Japan were at CAS talking up hypersonic aircraft—Burdick announced that the USDA had agreed to fund the ESSI building. In its rationale, it cited the planned study of—among other things— famine and drought and the use of satellites to help farmers forecast weather. In fact, it threw in $1.5 million for groundwater research.

Grateful for Burdick's intercession and eager to show his appreciation, Odegard invited the Democrat to give the speech at the dedication of the third CAS building—the one soon to receive the simulators from Northwest Airlines. It was an awkward decision, for that was a building whose federal funding had been secured by Republican Senator Mark Andrews before his 1986 defeat. The slight did not go unnoticed.

"John had a sugar daddy in Washington, Mark Andrews," recalls Kent Alm. "He and John pretty much agreed they'd put up a building a year. It was not to John's credit that instead of asking Andrews to give the dedication speech, he asked Burdick. I know Andrews sent a letter saying, 'So soon you forget, John.'"

Still, these were heady times for CAS. Tom Nunn, the Northwest executive who headed NATCO told the *Minot Daily News* that the University of North Dakota had become "The Harvard of the Air." The dropping of the H-word was pure icing on Odegard's cake. In the summer, a group of his friends had come together under the auspices of the local chapter of the American Diabetes Association to stage a roast in Odegard's honor. Among the guests doing the needling that night was Joe Lapensky, then chair of Northwest Airlines. He poked fun at Odegard's controversial reputation for being gone from the office so often.

 "John has been known to hang around North Dakota for three days at a time," Lapensky said, to much laughter. And recognizing Odegard's inability to stop his endless schmoozing, he noted that "John invented the five-hour banquet. The only banquet I've ever been to where everyone in the room was introduced." Lapensky then suggested that the state change its name to "North Dakotagard."

Don Smith told jokes about Odegard's expectations of an 80-hour-work week, to great laughter. When Odegard spoke at the end of the evening, he attempted to return fire at Smith by characterizing him as a man who showed the dean no respect. "Don says of me," Odegard quipped, "that 'Nothing is impossible for the man who doesn't have to do it himself.'"

Actually, that jibe said more about his own management style, particularly when it came to the unpleasant task of firing an employee.

"It was difficult for John to let anyone go," says Warren Jensen, M.D., a professor and the college's FAA medical examiner. "He had such high hopes for everyone. It was hard for him to accept that it didn't work out, and we don't want you in our organization anymore."

On the other hand, given Odegard's personality, it was inevitable there would be personnel clashes that would lead to a parting of the ways. In such cases, Odegard turned to the one man he knew would have no trouble with the odious task.

"He'd come into me," says Tom Clifford, "and he'd say, 'Could you get rid of this guy?' I'd say, 'Aren't you going to?' But he wouldn't do it. So he'd leave town and I'd do it."

There were also cases that year, just after the Chinese students arrived, where personnel clashes led two key people in the school to leave. In December—just three months after the success of his hypersonic conference—David Webb announced his resignation. According to other executives at CAS, a personality conflict between Odegard and Webb had been long simmering. It came to a head over the location of the proposed ESSI building. Webb had agreed with William Dando, chair of the department of geography, that the new building shouldn't be a part of CAS and should be built on the east side of campus.

According to the *Grand Forks Herald*, Webb had "strongly advocated a multidisciplinary approach involving several departments at UND besides those in CAS." The paper further reported that "Webb said he felt strongly that the planned earth science system center should be removed from CAS and placed directly under the president's office as a separate institution. That was the best way to emphasize the multi-disciplinary nature of the project involving several departments of the university."

At that point, though many of Odegard's scientists had earned themselves national reputations in their field, Webb was the superstar on the faculty. His resignation led to a headline a few days later in the *Herald*: "Hint All Not Well at CAS."

In the article, reporter Steve Schmidt wrote "...for all its glamour and growth, the aerospace center could be hiding some high stress points." It attributed to Webb the idea that "the aviation oriented environment at CAS could make it more difficult to organize an interdisciplinary effort of other campus organizations." It then quoted a space studies graduate assistant, John Muncy, whose dismay at losing Webb was evident. "This state," he said, "must have a death wish."

While Webb's departure raised eyebrows, a second personnel disruption bore a more personal impact. Don Smith, who to this day refers to Odegard "as almost like a son," left after 17 years of largely volunteer employment. Publicly he resigned his post as associate dean to become the executive director of the Aerospace Foundation. Smith won't discuss the topic, but several associates of his and Odegard say the two began growing apart during the startup of the China Airlines project.

"There was friction," says Bob Muhs. "It got very pointed. Don was a marvelous gentleman. He was an entrepreneur before anyone knew what the word was. He provided information and knowledge that John didn't have. I think John outgrew it after awhile, quite frankly. Some people reached the limit of what they could contribute to him."

While it would seem unlikely, if not impossible, to immediately find someone of Smith's expertise and stature to assume his duties, Odegard's famous luck beat the odds once more. Among those who saw his *Chronicle for Higher Education* ad for a new associate dean was retired 2-star Admiral Robert H. Shumaker. His academic credentials were top of the line. He'd graduated in the top one per cent of his 1956 Naval Academy graduating class. During a 30-year-career, he'd earned a Ph.D. in engineering and had been president of the Naval Post Graduate college in California. He'd been the final man cut from NASA's whittling down of 9,000 candidates to the 14 astronaut trainees it selected in the seventies when the space program was rejuvenated. During that grueling testing, Shumaker's roommate had been Buzz Aldrin who, in spite of his later North Dakota connection, had nothing to do with Shumaker's becoming aware of UND.

At the moment, Shumaker was an associate dean at The George Washington University in the nation's capital. During a business trip, Shumaker stopped over in Grand Forks for a bit of reconnaissance. When he did, it wasn't his academic heralds that caught John Odegard's attention. Shumaker's rise in the ranks to admiral and to academic stature came only after he'd returned home in 1973 following eight years as a prisoner of war in North Vietnam.

He was the Navy pilot, shot down over North Vietnam in 1965, who coined the name "Hanoi Hilton." It referred to the notorious Hoa Lo prison in the North Vietnamese capital. During his eight years of captivity—three of them spent in solitary confinement in a dank, windowless room in leg irons—Shumaker endured brutal torture at the hands of his jailers. But it didn't keep him from resisting. He developed an elaborate code system by which he tapped out, on prison walls, reassuring messages of hope and courage to fellow prisoners he couldn't even see. One of his cellmates at Hoa Lo was John McCain, now the U.S. Senator from Arizona. When he was released in 1973, Shumaker was awarded two silver stars and the Navy's Distinguished Service Medal for his heroic resistance and help to his comrades.

Odegard liked Shumaker and took him before his selection committee for an interview. At that meeting, one of the committee members, learning that Shumaker had been a POW, recalled that he'd recently heard an inspirational talk by a man who'd also been a North Vietnamese POW. This man told of being shot down and dragged through the streets and thrown into a prison cell that was nothing more than a corrugated iron box. After awhile he heard a scratching sound. Then someone pushed a wire into the cell with a note attached. It said, "You're not alone. Here to help you." That, said the member of the selection committee, had been the epiphany for this POW. It had turned him around. Turning to Shumaker, he asked if he'd known that POW. Shumaker remained silent for several moments and then in a choked voice said, "Yeah, I wrote the note."

And so, he and his wife Lorraine, charmed by Odegard and applauded by the committee, moved to North Dakota. With David Webb's resignation, Shumaker not only became the new associate dean, he also assumed command of the space studies graduate program and got its anxious students back on track.

In 1989 Odegard hired retired admiral Robert H.
Shumaker (center), a prisoner of war in North Vietnam
for eight years and the man who coined the term Hanoi
Hilton. Shumaker (here with his wife Lorraine) became
associate dean and introduced Odegard to another
POW friend, United States Senator John McCain.

John Odegard, George Hammond and a guide pose with their walleye catch of the day
during an outing to the fishing retreat in Oak Lake, Manitoba. It was a favorite spot for
Odegard to host VIP's from the political and aviation world.

Shumaker found Odegard refreshing, but like no one he'd ever worked with before. "I appreciated John and bonded with him and think the world of him to this day," he says. "I did get a little nervous about some of his entrepreneurial ventures. Some did well. Some went bust. But those minor negative things are really things I kind of admire him for."

As a typically conservative military planner, Shumaker says he came to Grand Forks prepared to focus on the bottom line. Instead he found that Odegard seemed to have misplaced it.

"It was the difference in our personalities," he says. "I'm the guy in the back of the cockpit, the engineer. Like in the Jonathan Winters comedy routine. Winters played Speed Davis, the cocky test pilot. I was the engineer saying, 'It's not going to fly, it's not going to fly.' That was kind of our relationship. I was always trying to hold down costs. I probably would have retired with three stars, but I'd be the guy at the Pentagon who, when they're talking of spending multi-millions on something, I'd say, 'But have you considered this...' John would have said, 'Oh, man, go for it.'"

Shumaker wore many hats as associate dean. When CAS landed a contract to train students from Bahrain and the United Arab Emirates, those students were initially placed in regular campus dormitories. When one of them decided to slaughter a sheep in the shower room, it fell to Shumaker to tactfully resolve the matter. He also became adept at balancing the needs and expectations of two disparate groups: the young and coming aviation executives who had started early with Odegard, and the non-aviation scientists and academics who often saw things in a different light but sometimes had trouble reaching the dean.

"I fancied myself as a go between for those two groups," says Shumaker. "A lot of them, who couldn't get an audience with John, laid their stuff in front of me. I don't know whether John felt intimidated by them or his interests were just more in aviation. But I would then present their petitions to him."

Shumaker became the face of CAS to the rest of the university. Odegard also assigned him the role Don Smith had played so well: attending the monthly meetings of the university's deans and taking the heat they usually supplied. "Most of them thought we were kind of the back door people," says Shumaker. "They thought our academic standards were not as high as theirs."

On one of their trips to Washington, Shumaker took Odegard to John McCain's Senate office and introduced them. They took an instant shine to each other, and it wasn't long before Odegard had invited McCain to one of his famous fishing junkets at the Oak Lake resort in Canada. Shumaker and Bob Buley went along on that trip which generated one of the more famous Odegard good-bad boy flying stories.

"One day up there," says Buley, "John and I decided to take the Cessna out and go trout fishing on another lake. We left the rest of the crew in camp. Shumaker and McCain decided to hike up to a little lake where there were some nice big bass. So we fire up the plane. We knew McCain and Shumaker were up there. John spots them out in their 14-foot aluminum row boat on this lake. John says, 'Let's buzz 'em.' Shumaker sees us coming, and he bends over and moons us. John says, 'The son of a bitch is mooning me.' I said, 'John!' and I grabbed the stick, and we just missed the trees."

Buley recalls that McCain later told everyone that before Shumaker mooned Odegard, he'd set his rod down in the boat. While the admiral was doing the mooning, his rod started to go over the side. McCain grabbed for it just in time. When he pulled it in, he found a heavy, wiggling bass on the other end.

> The public demands certainties...
> But there are no certainties.
> H. L. Mencken

Chapter 21

Some Disappointed People

In those early days of the sixties, when John Odegard was delivering his Flying Club recruitment pitch on the run in the student center, it would have been difficult to imagine a time when *too many* people decided to take him up on his offer. But in 1989, the Center for Aerospace Studies numbered 2,300 majors in its four departments—1,854 of them in aviation. One out of every six students at UND that fall was matriculating west of the Coulee. Three CAS buildings were in place on campus, a fourth was just getting underway and a five-story tower nearing completion at the airport. On paper, not even John Odegard could ask for more.

In just one year, his total enrollment had increased by 50 per cent; students were coming in from all over the world. The place was "bulging at the seams," he told the *Grand Forks Herald*. While one can imagine such a remark being made with a smile and a hint of pride, there was actually a sense of panic at the school. For it wasn't just bulging. There were simply too many students for the facilities, the faculty and the fleet to handle.

"Those were wild times," says Ken Polovitz. "We brought in 650 new students in one year. A lot of them had to wait until their second year to start flying. I remember we were filling out the classes very fast, and we had five slots left in the private pilot section. We made this big announcement that we would give them out on a first-come first-served basis, starting at 9 a.m. Students were camped outside in sleeping bags, lining the hallways. John came to work and had to step over these kids to get into the building. I said to him this is the result of all the hype we've been doing. But he loved it. Of course, he didn't have to manage it. He turned to Kent Lovelace and me and said, "You guys take care of this.""

Easier said than done. The immediate remedy was to try to console those freshmen who lost out on the chance to fly right away. "It didn't hurt these kids academically not to fly right away," says Polovitz. "We could put those kids in classes, working on their four-year-degree. But most said, 'We came to UND to fly, and if we can't start right away, we're unhappy."

Parents who were footing the hefty bills for their sons and daughters let Odegard know about their displeasure. But even students who did get to fly would sometimes show up for a scheduled flying lesson only to find there were no aircraft available. Actually, there *were* planes sitting on the ramp, but they were reserved for China Airlines students. The way things worked then, CAS owned a fleet of planes, and the Aerospace Foundation leased a separate fleet for the China Airlines program. The two groups of planes were scheduled separately. The regular undergrads began to feel unappreciated and, again, made their feelings known.

Odegard could only explain to disillusioned students and their parents that more planes, more simulators, more facilities, more efficient management tools were on the way. He had solid reasons to be optimistic. After years of trying, he'd finally made headway with the state legislature. In 1990, it agreed to appropriate funds to CAS, setting aside $200,000 as seed money to encourage ventures with IBM and Control Data in developing and marketing software for flight operations and scheduling. Of course there was also the nicely humming Spectrum program, with clients such as Gulf Air in Bahrain and the Arabian-American Oil Company—ARAMCO—sending students to Grand Forks. Even the French Airbus company had sent scouts to Grand Forks.

Meanwhile, the "hype" that contributed to the enrollment overload stemmed not only from effervescent Odegardian tub thumping. Ever since 1986, a relentless beating of the pilot-shortage drum had rumbled from several industry sources. The prediction had it that hundreds, if not thousands, of lucrative airline pilot openings would soon come available.

Another contributing factor to the overload was the low academic entrance requirement for new CAS students. There had been a time when Odegard was delighted to take anybody who wanted into the program. That gradually narrowed over the first 20 years, into a requirement of a 2.0 grade point average—essentially a C. But now, he told the *Herald*, "We're going to have to find some ways of screening out our weaker students."

As a one-time weaker student himself, the bulge in enrollment was a rock of irony pushing him up against the hard spot of success. Yet entrance requirements began to rise, at first from 2.0 to 2.3. Since about 1992, students have been required to maintain a 2.5 index to stay in school.

The school had gone through a similar burst of enrollment in 1979—followed in 1981 by an economic speed bump. It had forced Odegard's first layoffs. But the eighties had quickly bounced back to become the decade of excess, pushing everything once more back to the tippy top. It has ever been thus in the aviation industry: growth charts whose peaks and valleys resemble the outline of the Himalayas. Though no one seems to have seen it coming, history, at least was warning in the early nineties that it was valley time once more.

Economic analysts, puffing their pipes in some comfy aerie far from the madding crowd, are expert at euphemism. What they delicately call a "contraction of the market," the rest of the world knows as a rug-pulling recession. Down amid the debris of their terra infirma, some even call it a disaster.

In the summer of 1990, Saddam Hussein invaded Kuwait. Almost simultaneously, the stock market contracted—much in the way a boa constrictor shakes hands with a rabbit. Dragging well into 1992, the recession was the most severe stunting of American economic growth since the Great Depression. While it hurt all forms of business—about two million people were temporarily or permanently laid off—it was particularly unkind to the world of aviation.

Certainly, overcrowding was no longer a pressing issue to John Odegard. Only two years after his boom, CAS enrollment had fallen to 1,600. For many who had dropped out or transferred, it had become too costly or too frustrating to continue.

In the meantime, Northwest Airlines found itself up to its tray tables in the goo of hypersonic fuel prices, falling ridership among constricted passengers, and lots of well-intentioned but suddenly ridiculous-looking debt. It brushed up dangerously close to the dreaded b-word which euphemistic economists are loathe to pronounce because it's what happens when you go bust in Monopoly.

While the recession would ultimately exact a deadly toll on the Northwest-UND relationship, the seeds of trouble had been sewn years earlier. When Joe Lapensky, chair of Northwest Airlines, stood at the podium that night in 1988 during the jolly John Odegard roast, he dropped a revealing hint of the trouble. It came in the form of a throw-away line, an oblique reference to the simmering personality conflict between Odegard and Tom Nunn, the man Northwest had put in charge of NATCO, its training arm.

"Being between Tom Nunn and John Odegard," quipped Lapensky, "is like being between two bees in a quart jar."

The friction that had developed between the two men centered on which partner had ultimate control of the Spectrum program: UND or NATCO.

"John had a pretty big ego," says Bob Buley. "If you look at the relationships he's had with people he's brought in, at some point something would occur where a person is going to have to declare himself: either you're the number one guy or the number two guy. John was not about to let anybody but him run that program."

Tom Nunn, however, was known also as a strong manager. Yet the disagreements, to which Lapensky often referred, became moot in June of 1989. Private investors Alfred Checchi and Gary Wilson leveraged a buyout of Northwest Airlines for $3.6 billion. The dominoes began falling almost immediately.

The industry trend at the time was toward consolidation, with stronger airlines buying out the weaker—and there were plenty of the weaker variety. Northwest's new owners talked of buying up rights to failing Eastern Airline's Washington, D.C. operations as well as troubled Continental and the Australian carrier Qantas. Yet, only a month after Checchi and Wilson bought the airline, they decided to end Northwest's vaunted Spectrum investment at UND. The media spin made it seem like a natural development that the Aerospace Foundation was now buying out NATCO's interest in the entire *ab initio* training program. In truth, the failure of the partnership with the airline was a double blow for Odegard and his school.

Not only did Northwest withdraw, but the rationale that had made the NWA-UND connection feasible in the first place fizzled. It's hard to say exactly when the reality began to sink in—perhaps when Eastern or Pan Am folded—but in 1989 and 1990, industry people began to wonder whatever happened to the big pilot shortage. The need that had been predicted for so long—and on the strength of which Northwest bought into and helped launch the innovative Spectrum program—had simply evaporated.

The reasons really were very simple. When those weaker airlines went bankrupt or cut workforce to save money, they released a significant number of trained pilots into the hiring pool. These were pilots with 5,000 to 10,000 hours of experience—compared to the 500 hours a UND student then accumulated upon graduation. At the same time, according to Lapensky, an unexpectedly large number of military pilots had mustered out of the service.

"They thought there would be a large shortage," says Northwest executive Bob Muhs. "But the industry consolidated, and we had a recession. It's the downside of the airline business: economic cycles were such that a perfect storm came together. In the end, the pilot need wasn't there."

Ken Polovitz likes to return to the word "hype" when discussing that phantom shortage.

"My analysis," he says, "is that some middle managers and some directors at Northwest and UND—spearheaded by John—were *going* to have a pilot shortage. We needed avenues to feed our UND grads to Northwest. As the economy slowed down and was not expanding, the airlines retrenched. Every ten years it seems to do that. People realized this shortage wasn't as real as everyone said it would be. There were some disappointed people."

A year later in 1990, when the recession hit, UND's connection with Northwest eroded further. The airline's ambitious expansion plans, already saddled by an outstanding $4 billion in debt carried over from its 1986 purchase of Republic Airlines, made a forced landing. In 1990, the airline lost $500 million. Three other airlines in addition to Pan Am and Eastern—America West, TWA, and Continental—either ceased to be or were absorbed by others during the recession. Many of them imploded under debt accumulated by overzealous expansion during the eighties. Northwest came very close to bankruptcy, and while it survived, it surrendered the initiative in the airline consolidation game to its healthier competitors.

That meant that ultimately, the North Dakota-Northwest relationship became less and less important to the airline. "There was a need for our connection to UND at one time," says Lapensky, "but the need kind of went away."

It quite literally went away in the summer of 1990 just as the recession hit. Though it had given up the Spectrum program, Northwest still had 12 employees and two simulators in the building they had helped to build on campus. Under a major cost-cutting effort, the airline pulled the plug on all of its operations in Grand Forks, recalling its employees and demanding the return of its simulators.

"When we finally decided we couldn't work together," says Kent Alm, "the meetings we had with Northwest were pretty tense."

In the process of pulling out, according to a *Grand Forks Herald* report, NATCO technicians appeared on campus one day and started dismantling its simulators. The chief of the university police department was called in and told the technicians to stop. They were told that they did not have permission to remove equipment from university property. So they stopped, and the relationship with the airline fell into an even more contentious phase.

At issue were the remaining 13 years on the lease of the CAS building that NATCO had signed, committing it to annual payments of about $500,000 for its share of the building's cost. After a month of negotiations, NATCO agreed to continue the payments, and the simulators were released to them.

But by 1993, with its half of the building still sitting empty, Northwest tried to talk its way out of the lease. Because Bob Buley knew UND, he was assigned to the negotiating team that went to Grand Forks to plead the case with John Odegard. This could be characterized as roughly the equivalent of hiring Willie Sutton as a bank guard.

"What I already knew," says Buley, "was that Joe Lapensky had sat down with John when we wrote all the initial agreements for that building. Joe wrote the lease. He has a reputation as an incredibly astute businessman. His words to John: This building is not always going to be owned by Northwest or the same people. So we're going to write this lease, so no one could get out of it. And lo and behold, that's what happened. Northwest put its best legal minds to the problem and they couldn't find a way out."

In Grand Forks, the wink-wink, nudge-nudging between Odie and Bobby was something to behold. "John and I enjoyed the hell out of it, because he knew he had a solid lease, and he didn't have to do anything. He also knew what I had in mind," says Buley.

During the early 90's Russia wouldn't allow Northwest Airlines to overfly its territory for a short cut to the orient because its air traffic controllers couldn't understand English commands. Odegard (here in Moscow) solved the problem by bringing the Russians to Grand Forks for a course in aviation English.

"We looked at where we were in our vision when we started this thing. What was missing was a way to have our students being pre-selected and given preferential treatment for employment. So we figured out a way to let Northwest get out of some of the lease by putting together a preferential hiring program. They went for it."

Or appeared to, anyway. Northwest staged a major press conference in Minneapolis in the fall of 1993 to announce its "Preferred Pilot Hiring Program." Yet it seemed to be doing a little wink-wink of its own. On the one hand, the airline said it would give first dibs on pilot jobs to UND Spectrum graduates. (By now, Odegard had made the *ab initio* curriculum available to qualified UND students and not simply to those underwritten by foreign governments or foreign airlines.) In response to questions, however, Northwest officials admitted that because there currently was "a pilot glut," they weren't hiring anybody at the moment or at any foreseeable moment down the road.

"The PPH program was a lot of PR," says Polovitz. "I think that's when I lost all of my hair. I got six calls a day, kids saying, 'Sign me up. I want to go directly to Northwest.' But no Spectrum grad ever was placed into Northwest."

The airline, he says, sloughed off the PPH program onto its regional carrier, Mesaba airlines, where 8 to 10 UND grads did find work. "When we pressed the issue with Northwest," says Polovitz, "the people we'd signed the agreements with weren't there any more. I'm not sure their vice presidents ever bought off on the idea—maybe the concept, but not the actual happening. They didn't need pilots."

"I don't think there ever was a likelihood," adds Kent Alm, "that Northwest would take our graduates and right away put 'em into one of their jets."

PR or not, the PPH program was only one of several considerations agreed to by the university and Northwest to assuage those lease payments. The university negotiated with the Bank of North Dakota—which held the note on the CAS building in question—to allow Northwest to pay just the interest on the lease for one year. Northwest also agreed to send business to the Aerospace Foundation when it came across opportunities. This actually led to tangible results when Bob Buley came up with a win-win idea for both parties.

For years, Northwest had been trying to get permission from Russia to fly over the country on its Asian routes. Avoiding the roundabout route to places such as Japan, Singapore and Hong Kong not only would save the airline time, but costly fuel. However, the airline had gotten nowhere with the Russians.

Buley and Odegard tried their hand at the negotiating table and charmed several Russian military officials. They learned that one of the barriers to allowing Northwest to overfly the country was that most Russian air traffic controllers didn't understand English—the standard language of international flight communications. Rather than try to teach the Russian ATCs a full course of English, Buley suggested

Ken Polovitz (left) the Assistant Dean for Student Services and Kent Lovelace, Chair of the Department of Aviation, often brainstormed solutions to suddenly burgeoning student enrollments, which just as quickly could become declining enrollments in the volatile job market of the aviation industry.

Visitors to The Odegard School are often amazed at the high-tech training center for air traffic control students. The unique program includes training on cutting-edge radar and tower ATC simulators.

that the Aerospace Foundation train them just in the English commands and responses they would normally use and hear during communications with an airliner.

The Russians liked the idea, as did the U.S. State Department. When it came to who would pay for that training, Buley had another idea. Northwest would have to pay Russia for over-flight privileges. In a three-way deal, the Russians agreed to forego some of the over-flight fees and have Northwest apply them to the Aerospace Foundation's training. In return, the foundation applied some of those funds against the remaining NWA balance of the CAS building lease.

The aerospace school eventually trained several dozen air traffic controllers from both Russia and the Republic of China in aviation English. In the meantime, the FAA, (as part of the Airway Science curriculum it had earlier funded) had approved a new major for CAS students in Air Traffic Control. The FAA was impressed that CAS had developed its own air traffic control simulator, the ATC-2000. Graduates would have the same ATC certification that students got at the FAA's own training facility in Oklahoma City. That marked the first time the FAA had allowed any but its own trainees to be so certified.

Meanwhile, the university devised a formula for Northwest in regard to its CAS lease obligations. It formalized a set of lease-offset percentages that would pay down the airline debt in any future business deals it referred to UND. Even so, a few years later, Northwest told Odegard it wanted once more to renegotiate an end to the lease. In the midst of those negotiations, Northwest surprised UND by suddenly bringing suit to void the lease. A compromise was eventually reached in 2000. UND agreed to pay about a third of the remaining $1.5 million owed on the lease. It settled the issue, gave the university full ownership of the building and effectively wrote the end to a relationship that once had held so much promise.

Accept the premise; you'll enjoy the bit.
David Letterman

Chapter 22
The Next Thing I Knew, I Owned an Airplane

Right about the time that those two bees were buzzing in Joe Lapensky's quart jar, another key outside relationship that Odegard had nurtured for years was in danger of coming apart. This time the controversy had nothing to do with Odegard's sometimes overbearing management style, nor was it to be saved by his always irresistible charm.

It involved a nearly charm-proof organization, the United States Army, and a grandmaster of the overbearing, Dick Cheney. Long before he was George W. Bush's vice president, Cheney had served George H.W. Bush as a cost-conscious Secretary of Defense. One day early in 1990, Cheney made headlines by publicly complaining about the expenditure of time and money on a program the Army had been financing at UND for almost eight years.

This was Air Battle Captain, the program dating to 1983 in which CAS trained Army ROTC cadets to fly helicopters. Each cadet graduated with a commercial helicopter pilot's license and then reported to the Army's helicopter flight school at Fort Rucker, Alabama. Odegard had sold the program to the Army as a cost-savings measure, on the theory that students who arrived at Fort Rucker already knowing how to fly a helicopter were bound to cost less to train and become better combat pilots than those with no experience at all. And, in fact, three out of every five UND grads who had reported thus far to Fort Rucker finished in the top ten of their class.

But under Cheney's watch, the Army decided to cancel the program to save money. It would have been a stunning blow to a program that had $2 million invested in its helicopter fleet. Pressed by Odegard, Senator Quentin Burdick got the Army to delay its cancellation while it conducted a study on the usefulness of the program. This prompted Cheney to complain to Congress that the expensive study had "absolutely nothing to do with the safety and security of the United States."

Odegard put a brave face on the criticism, responding that Cheney didn't have a problem with the program, per se, but with the paperwork involved in the study. The Army wasn't so sure. Odegard then turned the case over to retired Air Force Major General Darrol Schroeder.

The two had been friends ever since the sixties when Schroeder interceded with a group of fixed base operators who wanted Odegard's head. Schroeder, like Odegard, had been a crop sprayer. He started his own agribusiness when he left active duty in the Air Force and continued to fly with the Air National Guard in Fargo, eventually earning two stars before retiring in 1988. Odegard had persuaded him to come to the Aerospace Foundation in 1989 as a contract marketing specialist. He would later serve as the foundation's director of business development.

Schroeder, who knew the ways of the Pentagon, realized that cost wasn't the real issue behind the Army's wanting to kill Air Battle Captain.

"They were looking at it as competition," he says. "They thought why is a university training our helicopter pilots?"

Schroeder went to Alabama and found a receptive audience at Fort Rucker. "They knew it was a very cost effective program," says Schroeder. "We'd proved to them we could train those cadets cheaper in North Dakota than at Rucker."

His next visit was to the real stumbling block, the Pentagon and the Department of the Army. Schroeder brought with him a member of Senator Burdick's staff, a man who had once served as an assistant secretary of defense.

"But we weren't very well received," Schroeder recalls. "We met with a one-star general who kind of degraded us a little bit."

Burdick's aide, however, made it clear that the senator wanted the program. The Pentagon general ultimately relented and, after the "usefulness study" was completed, the program was reinstated.

Schroeder was also instrumental in returning an Air Force ROTC program to UND. It had ceased operation in the sixties and several times over the years the Air Force had resisted Odegard's efforts to reinstate it. An Air Force ROTC program existed in Fargo at UND's rival North Dakota State University and the Air Force thought that was enough.

Since Schroeder was an Air Force man, he took it as a personal mission to get the program back to Grand Forks. After months of meetings in Montgomery, Alabama, the headquarters of ROTC, Schroeder realized there was only one way it would happen. UND would have to become part of the NDSU's ROTC detachment.

The news didn't go over well in Grand Forks where Tom Clifford refused to

Retired Air Force General Darrol Schroeder's relationship with John Odegard dated to the early days of the aviation department program when he defended Odegard against grumblings by local fixed-base operators. Later, as an Aerospace Foundation marketing officer, it was Schroeder's idea to get an altitude chamber.

Odegard refurbished this surplus altitude chamber donated by the U.S. Air Force. It is used to teach future jet pilots how to recognize the deadly onset of oxygen deprivation in the event of a malfunction in a pressurized jet cabin.

sign the agreement. Since their founding, UND and NDSU—separated by only 70 miles—had been fierce athletic and academic rivals. Still, after more negotiating, the Air Force refused to budge. Odegard and Schroeder both felt the aerospace school could only benefit by having an association with the Air Force and, in fact, might appear to be diminished without it.

Reluctantly, Clifford agreed, but not without grumbling. Ever since, however, the UND ROTC detachment has grown much larger than the NDSU group, with its own full time staff in Grand Forks. Hardly anyone ever notices that its headquarters actually remains in Fargo.

In June of 1991, while Schroeder was at work on those adventures, retired Admiral Bob Shumaker left his position as associate dean. The official reason given was a difficulty adjusting to the cold weather of North Dakota. But that seemed odd. Shumaker had grown up in northeastern Pennsylvania and his wife, Lorraine, hailed from Montreal—neither an especially balmy locale in the winter.

"There was a deeper cause," says Shumaker. "I didn't feel that John and I worked hand in glove; we weren't a good fit. Some of it was my fault and some John's. John was a very kind guy to me. He sponsored me onto the board of directors of the Aerospace Foundation. But he was reluctant to let the strings loose."

Not only would Odegard reserve all decisions—large and small—unto himself, says Shumaker, but he was gone for extended periods, during which he felt he had no authority to make decisions.

"It was kind of awkward," he says. "At one point, after a couple years, I was getting kind of frustrated. I said to him, 'Why not do away with this associate dean thing, and let me be the chair of the aviation department?' You know, I don't think he recognized he was doing that."

Kent Alm isn't so sure. "John had a great need for the center stage," he says. "It was his program. He was the star. He knew what he had done and he didn't really want to see anybody get too far out on anything."

A project that captures some of the frustrations felt by Shumaker—but which turned out reasonably well—was the oh-so-Odegardian case of the free altitude chamber.

The story dates from 1989, shortly after Shumaker's arrival, when Odegard began recruiting the retired general Schroeder. He had invited Schroeder to Grand Forks and had walked him through the CAS operation. At the end of the tour Odegard asked his impressions of the place. Schroeder remembers suggesting a physiology program. He explained to Odegard that military pilots go through a phase of training unavailable to most civilian pilots. It's a program aimed at helping jet pilots who fly in pressurized cabins to recognize the symptoms of hypoxia. This deadly lack of oxygen usually stems from a system malfunction that causes a gradual loss of cabin air pressure.

The idea, said Schroeder is to acquaint a pilot with the physical symptoms of hypoxia in a controlled setting. Then, if the onset of those symptoms occurs in an actual flight, a pilot would recognize them in time to take corrective action. Those without the training usually think they would be able to tell if they were slipping into unconsciousness, Schroeder noted. But, in fact, many times the onset is so subtle that if the pilot doesn't realize what is happening and quickly connect to supplemental oxygen or descend to a lower altitude, it is too late to react.

In the military, pilot trainees are put into an altitude chamber which simulates the pressurized cabin atmosphere of a jet. Oxygen is slowly drained away from a student's oxygen mask and the symptoms of hypoxia slowly begin. Once the session is over, a student has a better idea of how his or her body feels and reacts to oxygen deprivation.

"The altitude chamber," Schroeder told Odegard, "is a very serious part of aviation."

Intrigued, Odegard barked, "Get me one!"

"So I got him one," says Schroeder. "After three months of phone calls, I found a surplus Air Force altitude chamber in Salt Lake City."

In the meantime, Odegard promoted the altitude chamber to his colleagues at CAS with a well-remembered phrase. "This chamber," he boasted, "will not cost us one cent."

In fact, the chamber Schroeder located had been cocooned for some time and would need refurbishing. To do that it would need to be shipped to its original manufacturer in Virginia and then shipped to North Dakota. Odegard went to the Aerospace Foundation and requested $20,000 for the refurbishing. Admiral Shumaker was immediately leery.

"I was a little bit hesitant about it," he says. "I didn't see the profit line on the thing. But John went ahead."

He did so because profit wasn't his motive. His thinking was that just having an altitude chamber would set UND apart from other aviation schools and serve as a magnet for students. The foundation board approved the cost. But at the next meeting, Odegard was back with a request for another $20,000, this to get the monstrous chamber from Salt Lake City to Virginia.

In a subsequent meeting, Odegard told the foundation board that safety regulations required a trained operator to be inside the chamber at all times with students. A crew of five people experienced in chamber technology would have to be hired. Again, the board agreed to the cost.

Don Smith, who was then heading the foundation, remembers not too long afterward when Bob Reis sent him an invoice for $160,000.

"I said, 'What's this for?' Well, it was for conditioning the altitude chamber. I said, 'My God we're not going to do this.' He said, 'It's been done.' I said, 'What?

Without a purchase order? We know nothing about it!'"

It turned out that Odegard had given carte blanche to an ex-Air Force officer he'd hired to run the chamber. "There was no legal position that we couldn't pay these people," says Smith. He remembers confronting Odegard, whose response was a matter-of-fact, "Yeah, we needed it."

But where would the money come from? Reis suggested financing the chamber the same way they paid for airplanes, with a loan using future student fees as collateral. However, they were turned down by their finance company—and not because of the cost or credit problems. It seems the altitude chamber was technically still owned by the Air Force. Until the endless reams of paper work were finished and UND actually owned the machine, they couldn't borrow the money. Odegard told Smith he'd simply ask China Airlines to pay $600,000 in its first Spectrum payment instead of $400,000. By then the ownership issue should be cleared up.

"And he did ask them," says a still astonished Smith. "And he got the money."

But it wasn't the end of the expense. Once they had a completely rebuilt altitude chamber they needed a place to put it. Odegard had figured he would put it in the empty simulator space in the building recently vacated by Northwest Airlines. But relations with the airline were then in a state of high tension, and permission to use the space was denied.

The only other possible space was a small garage in CAS I, where Odegard parked his car. Smith then got the architects Schoen and Kobetsky to redesign the space without charge. In all, according to Terri Clark, today's director of fiscal affairs for the foundation and for the school, $900,000 was spent to get the chamber that wouldn't cost one cent delivered to UND and useable for student training.

In spite of those costly hijinks, the altitude chamber has been a hit. First, as school officials will note, Embry-Riddle doesn't have one, a moral victory unto itself. Odegard made sure the world knew that by producing promotional videos about the physiology program. Billing: we have the only altitude chamber between Minneapolis and Seattle.

Secondly, the chamber is still used today for all pilot trainees—under the guidance of Warren Jensen, M.D., a physician who oversees the chamber and teaches courses in physiology. Jensen's is a fascinating tale of the kind of risk taking and reinvention that Odegard so often encouraged and fostered.

For eight years, Jensen practiced medicine in his home town of Cavalier, a small burg north of Grand Forks. During those years as a general practitioner, he did everything from performing surgeries to delivering 375 babies. As is often the case with small town doctors, he was known and revered by everyone.

But ever since he'd been 10-years-old and caught his first glimpse of a Boeing 727, Jensen had been captivated by airplanes. He once considered a career as a pilot

In the early 80s Odegard's Center for Aerospace Sciences introduced helicopter training through a contract with the U.S. Army and its ROTC program. The Odegard school also trains cadets from the U.S. Military Academy at West Point each summer in helicopter flying.

Dr. Warren Jensen was a general practitioner in a small North Dakota town when he was bitten by the flying bug. He became one of the country's very few specialists in aerospace medicine and today runs UND's altitude chamber, teaches physiology and is a certified FAA medical examiner. To left is George LaMora; at right is Joe Schalk.

rather than being a doctor. But because he wore glasses, he didn't think it would work. A pilot friend in Cavalier suggested to Dr. Jensen that he get a certification as an FAA medical examiner—mainly so the friend wouldn't have to go all the way to Grand Forks to get his annual flight physical. Jensen got the certification and began giving flight physicals. Soon, he started bartering with his patients: the cost of a physical in exchange for a flight lesson.

"The next thing I knew," says Jensen, "I owned an airplane."

When he got his pilot's license in 1987, he signed on as a flight surgeon with the 119th fighter wing of the Air National Guard in Fargo. It wasn't too long before he felt a gnawing sense of unfulfillment being a small town general practitioner.

"I wanted to do more in my life," he says. "I decided to change my life, and I looked around at different professions."

His research led him to the very new and niche specialty of aerospace medicine—at the time there were only a handful of certified practitioners in the country. When he found out that NASA funded a master's level program in aerospace medicine in Columbus, Ohio, Jensen left his practice for a year's residency training.

The *Grand Forks Herald* wrote a story about the impact of Jensen's leaving small-town Cavalier. One of those who read that story was John Odegard. He dropped Jensen a note, wishing him well and inviting him to stop by. Jensen did, just before he left for Columbus. During that first meeting, Odegard told Jensen he wanted to add aerospace medicine to his curriculum. Like so many visitors before him, Jensen was bowled over by Odegard's infectious enthusiasm.

"I was so motivated by John's attitude and energy," he recalls, "I went downstairs to the gift shop and bought a shirt. Now I was a fan. I called my wife, and she said, 'How did your meeting go?' And I said, 'I feel like I could dig the Panama Canal with a spoon all by myself.'"

A year later, officially certified in aerospace medicine, Jensen turned down a job offer from Odegard in favor of a prestigious position with NASA as a flight surgeon. But immediately, something didn't feel right. The more he thought about it, the more Jensen came back to that statement he'd made to his wife about digging the Panama Canal.

"I waited three days to call John," he says. "I was scared. I didn't want him to think I was a nut. But I called him and said, 'I don't think I'm going to be happy at NASA. This fits better with me.'"

Odegard's response reassured him: "I'm even more fired up than before." He gave Jensen plenty of room to invent his job at CAS. Eventually the doctor took over the storied altitude chamber and devised a curriculum in the space studies program built around classes in human factors and flight physiology.

"Regular doctors take sick and injured people and make them better," says Jensen. "I start out with normal people to begin with. I subject them to hostile environments and get them to perform in spite of that. I look at peak human performance in spite of adversity. Most docs talk about treatment. I talk about countermeasures."

His approach to the altitude chamber is to teach students not only how to recognize their limits but how to optimize their performance. "I give them the experience so they can see it coming. And then I give them the counter measures of education, appreciation of risks, understanding how equipment works and how to take care of it—and then to have the confidence that the equipment will solve the problem."

Sometimes his medical colleagues in town will ask him what the heck he does over there west of the Coulee.

"I'm the only one in the world I know of who does what I do," says Jensen. "A medical doctor on the faculty of an aviation school? I know there are people whose lives I saved when I was in Cavalier, but I've never done anything before as exciting as this."

> What kind of man would live where there is no daring?
> I don't believe in taking foolish chances,
> but nothing can be accomplished
> without taking any chance at all.
> Charles A. Lindbergh
> Paris, 1927

Chapter 23
Who *Wouldn't* Want a Super Computer?

Bill Shea, the former FAA official who served for a time as chair of the aviation department for John Odegard, says he saw only one weakness in his boss.

"But it was a good weakness," says Shea. "He might have 10 balls going at once. He'd keep 'em going, and there might be a moment where he knew he could close a deal over here, but he was tied in over there. Put the two of us together in a room for 20 minutes and 40 different subjects would come up. What about this, what about that. He always had incredibly large projects going on."

Earl Strinden saw the same thing and sometimes worried about it. "There are always times with someone like John," he says, "when you can move too fast. You get so many balls in the air it becomes difficult to keep them all up."

But like those daredevil jugglers who can somehow keep a bowling ball, buzzing chain saw and napkin ring constantly shifting through the air, John Odegard held to the juggler's code: never stop moving. In those early, good-news-bad-news years of the nineties, CAS was still cutting a decent swath through the clouds. Its combined structures and fleet were worth an estimated $63 million. It boasted 700 employees and had a dozen simulators. Odegard could juggle all of this and dash confidently between an erratic economy, the divorce from ex-pal Northwest, and Dick Cheney's sarcasm, because he knew he had a solid gold lightning rod: It was called China Airlines.

That very first group of students from China Airlines in the fall of 1988 included flight engineers and flight attendants. Among the dignitaries applauding at their graduation in the spring of 1990 were General Shan, North Dakota

Among the first class of China Airlines students who trained
in North Dakota to become pilots was Liu, Yen-Ling, (left)
known to all as Linda. Pictured here with fellow student Lee,
Cheng-Kang, Linda Liu went on to become China Airline's
first female captain.

These six China Airline students were the first graduates of the expanded Jet Spectrum
training program at North Dakota. CAL purchased two Beechcraft jets for the program,
including this Beech 400 whose glass cockpit instruments are the same as those on a 747-400.

Governor George Sinner and U.S. Senator Kent Conrad. There had been only one wash-out among the first 24 trainees, and two of the Spectrum graduates—Linda Liu and Pam Chens—went on to become the first female pilots in Taiwan's history.

While at least those first 24 students had some experience with an airline, later classes included candidates such as doctors and college professors with no flying experience at all. The original Spectrum program gave students 1,000 hours of ground school training and 300 hours of pilot training in single-engine Piper Arrows and twin-engine Piper Seminoles. When graduates went back to Taiwan they became junior members of a passenger jet's cockpit crew with the expectation that one day they would become captains.

Almost immediately, however, veteran China Airlines pilots in Taiwan—many had earned their stripes as military pilots in the nationalist Chinese Air Force—mistrusted their new crew members. Specifically, they noted that Spectrum graduates had only 300 hours of flight experience—none of it in jet or turbine powered aircraft—when the requirements in Taiwan for a senior commercial pilot's license was 750 hours.

In the fall of 1990, the airline asked UND for an expansion of the training to 750 hours. It even offered to supply brand new turboprop aircraft for the purpose. Thus, in 1991, UND Aerospace began training close to 200 students in an advanced version of Spectrum. Students from that point on would receive the full 750 hours of training required for licensure in Taiwan. Through its U.S. subsidiary, CAL-Dynasty International, China Airlines spent $18 million to obtain four turbine engine aircraft: a pair of Beechcraft King Air C-90s and a pair of larger Beechcraft 1900s.

Advanced Spectrum Students began their Grand Forks training as usual in the single-engine piston-drive airplanes, progressing to jet-powered turboprops. The advanced program meant an extra $75,000 per student with the first advanced contract worth $13 million to UND.

"That put us on the international stage," says Al Palmer.

Since the fall of 1988, groups of students from international clients including Gulf Air, Aramco, Saudia and EVA airlines had been arriving in Grand Forks roughly every 90 days for an 18- month training period.

When China Airlines offered to buy the turbo props needed for the Advanced Spectrum program, John Odegard knew just the man to handle the sale. For as fate and John Odegard had ordained it, Jim Bunke, his erstwhile and earnest protégé, had taken his UND diploma and gone to work in the sales offices of Beechcraft in Wichita, Kansas. For a boy who'd always wanted to be an airline pilot, one might find this a bit of a grounding. While at UND, Bunke himself had loudly dismissed any suggestion he would take any job in aviation but airline pilot.

Then one day in his senior year, Odegard stopped him in a hallway and told him a recruiting team from Beechcraft was on campus for a week. They were looking for sales and marketing people. He told Bunke he'd set up a series of job interviews for him with Beechcraft over the next few days.

"I said, 'Well thanks, John. It's nice of you to think of me but I'm going to be an airline pilot," Bunke recalls. "But John says, 'No, no, this appointment is Monday and it's all set.' I said, 'Well thanks again. I don't want to waste their time or my time for that matter.' And he got right in my face. And he said, 'You're not hearing me well. You *will* be there at 8 a.m. Monday morning.'"

At the time, in the mid seventies, jobs for young pilots at airlines were scarce. Bunke hadn't worried about that until Odegard stopped him. But he proceeded to give it a lot more thought before he went into his first of several Beechcraft interviews on campus.

After the first job interview that week, Odegard debriefed the Beechcraft representative on Bunke's chances. The next time he saw Bunke he simply said, "Work on eye contact."

"So in the next interview," says Bunke, "I stared the guy down the whole time. And John gave me the feedback: the eye contact was good."

Good enough that two weeks after Bunke graduated from UND in 1978, he was selling airplanes for Beechcraft in Wichita. After two decades, he became the national sales director of Bombardier business aircraft in Minneapolis.

"The upshot," he says, "is that after 27 years I'm doing exactly what I interviewed for. I can say John did know better than I did what I'd be good at and what I'd like as my life's career. I interviewed for five days with the Beechcraft rep and everyday it became clearer to me that I was more suited to aircraft sales and the business side of general aviation than the airline world."

So it was natural for Odegard to pass the word to Bunke—who was handling sales for the Beechcraft King Air C-90 at the time—about the need of China Airlines and CAL-Dynasty for some turbo props. While Bunke immediately got his foot in the door and was looking at two quick sales of C-90s he remembers having a hard time convincing his bosses at Beechcraft.

"I'm going to the CEO and the vice president for sales and saying, 'I think I'm getting close to a deal for selling these planes.' And there's this look in their eye. Like, 'Now tell me again, this school is where? In North Dakota? Oh, really. And, you're going to sell them how many…?'"

He still winces at the memory of a sales trip he took to Grand Forks in the company of his VP for sales and the president of the company. "The VP pulls me aside," says Bunke. "He had tickets to be at the Master's golf championship, and we're in Grand Forks, North Dakota, and it's windy, and we're pitching a deal. He says to me, 'You son of a bitch, you got me up here.' It was kind of like, 'Who are we trying to kid? We're trying to sell these planes, and I'm missing the Master's.'"

But sell them, he did. Bunke remembers well the trip he took to Taipei to close his deal. Odegard had briefed him at length on the social dos and don'ts of Taiwan, emphasizing that under no circumstances should Bunke ever mention "the other China."

"He was teaching me the Taiwan traditions, what they eat, how they negotiate," he recalls. "I was in charge of this luncheon we were having. We had the Chinese flag set up. As the dignitaries are coming down hall, John pops in to see where we're all sitting. And he looks at the flag and goes, 'Bunke! Goddammit, that's the wrong China!' I turned white. Then John says, 'Just kidding.' He was fun that way. There was a lot riding on this for him, but it was always fun."

In addition to buying the pair of BE C-90s from Bunke—turboprops that could hold up to six people, including the crew—CAL Dynasty procured on its own two BE 1900s. This larger version of a turboprop seated 21 including the crew. It quickly became the standard craft of small, regional airlines.

The 1900s were brought in to give the China Airlines students a more realistic feel for flying commercial passenger planes. In fact, to make it completely realistic, the Aerospace Foundation and China Airlines essentially started an airline in North Dakota.

They advertised for bids among regional airlines, looking for one that would extend its routes into North Dakota. The winner of the bid was Great Lakes Aviation, a commuter airline out of Spencer, Iowa. By leasing one of the BE 1900s from CAL-Dynasty, Great Lakes was able to offer service from Bismarck, Jamestown, Fargo and Devil's Lake to Grand Forks and St. Paul. Great Lakes later partnered with United Airlines to operate some of its United Express regional routes, adding four daily flights between Grand Forks and Denver.

The pilots of those Great Lakes flights were all seasoned airline veterans. The co-pilots were China Airlines trainees. The operation became the ultimate flying laboratory, with regular UND Aerospace students also getting training in airline operations and management through the setup. Once the Taiwanese students had completed their training, they were accepted back in Taiwan as legitimate junior crew members. There, it became the job of senior airline pilots to further initiate their co-pilots in the unique skill set required to operate one of China Airline's swept-wing passenger jets such as a 737, 747 or Airbus 300.

The program worked very smoothly on both ends for three years. But in April of 1994, a China Airlines A300-6006 Airbus crashed on landing in Nagoya, Japan, killing 264 passengers. It turned out that the senior captain of that flight had ordered his co-pilot, a recent UND Spectrum grad, to land the plane. The problem: the experience gleaned from flying the straight wing BE 1900 in North Dakota was radically different than suddenly trying to fly an Airbus with its swept wing design. It was a skill that the senior captain had not yet taught his young co-pilot.

It was one of a series of China Airlines accidents, none as deadly as the Nagoya

During one of his trips to Taipei, Taiwan, Odegard joined in a night of karaoke singing with Diane, Rick Mercil, Al Palmer, Bob Reis, and Captain James Yu of China Airlines.

In this aerial shot of the Odegard aerospace campus in the late 90s, all four CAS buildings are in place. In 2002 a Hilton Garden Inn was built on the bare spot in the center of the picture. A $1 million walkway today links Ryan Hall, (top right) with Clifford Hall, carrying students safely above railroad tracks and busy 42nd street.

crash. In the others, Spectrum grads had been in the cockpit, but none had been the pilot. Odegard was deeply troubled by all of them and immediately wanted to find out what had happened and how such accidents could be prevented.

According to Bob Reis, Odegard got hold of cockpit tapes from those flights and had them analyzed. "It was obvious our students knew what was going on," says Reis. "But the senior officers on those crews were not taking their junior officer's advice on how to get their plane under control."

Odegard knew what was needed: a program that took all *ab initio* students through a transitional phase in actual jet aircraft. The course would familiarize students with the new so-called glass cockpits, faster planes, the idea of getting into air space quicker, and landing a plane quicker.

Odegard also suggested that China Airlines send its senior officers through the same training as their junior crew members. The airline agreed and thus began the third phase of the program—Jet Spectrum. For this segment, CAL-Dynasty made Jim Bunke's career by purchasing a pair of 8-passenger business jets through him—the Beechcraft 400. Its glass cockpit instruments were the same as those on a 747-400.

Starting in 1995, students from the airline spent an extra 90 days in Grand Forks flying the BE400s and working specifically on takeoffs and landings. All previous Spectrum grads were brought back for the same jet training.

"Each student now stayed with us for two-and-a-half to three years," says Palmer. "They started off in a little single engine airplane and ended up flying a jet. Just imagine: This program required a marriage between China Airlines, Great Lakes Aviation, UND, and it also had the blessing of the Airline Pilots Association. Only Odegard could have pulled it off."

By the time the China Airlines program phased out in 1997, the UND Aerospace Foundation had taken in more than $55 million. Managing those China Airline contracts required a Niagara of energy for the almost daily nurturing, charming and troubleshooting on behalf of a very particular client. In all of it, John Odegard was the man China Airlines wanted to deal with. Though the pace was often frantic, the experience led to a rounded sense of well being that John Odegard hadn't often felt.

"John was very happy and up after the China Airlines deal got in place," says Diane Odegard. "I saw him feel more comfortable than ever before. He was in big demand for speaking, he had great international recognition. It was smooth sailing here—not all the time, but pretty nice."

The Odegards could now afford to remodel their Reeves Drive home. John would get up early and bike ride around town, take a steam bath in his basement spa, wrap a big, thick robe around himself and settle down in the family breakfast nook to a café au lait.

"He started to appreciate the comforts of home," says Diane. "This was a high point. Things were going well."

China Airlines, however, was only one ball among many being held aloft by the force of Odegard's will. A couple of them got heavy enough that a fumble seemed almost certain.

For example, in the early part of 1992, forecasts for job growth in the aerospace industry were dismal at best. The recession was slowly coming to an end, but it would be awhile before the industry caught up. Among the aggravating factors for the aerospace world was the breakup of the Soviet Union and the end of the brief Gulf War—prompting the Pentagon to consider a cut of about $72 billion in its budget. Military contractors, many in the aviation industry, were predicting the loss of thousands of jobs. The Gulf War had also disrupted the flow of students between UND and Bahrain and Gulf Air. That meant a loss of revenue that put a greater strain on the flying fees of regular CAS undergrads—a key source of CAS income. Enrollment still hadn't rebounded from the drop a year earlier, meaning the fees they did collect were short of projections. Meanwhile, for the 200 CAS seniors about to enter the job market that spring of 1992, the prospects of finding a job in their field hadn't been as bad since the seventies. Ken Polovitz, the director of student services, told a reporter, "The market is flooded with people."

Even the huge purchase of planes from Piper four years earlier was coming back to bite the program. Piper had fallen victim to bankruptcy during the recession. While it was eventually able to supply CAS with the planes it had promised them, Odegard in the meantime, was forced to buy needed single engine craft from Aerospatiale General Aviation in France at a cost of almost $270,000.

Worse, the price of suddenly hard-to-find spare parts for the Piper planes in the fleet had exploded by 1,000 per cent. "We're getting killed on Piper aircraft parts," Odegard told Steve Schmidt of the *Grand Forks Herald*.

And although Quentin Burdick and Darrol Schroeder would eventually talk the Army into continuing to pay UND for helicopter training, the 24-month suspension of the annual $2 million funding for Air Battle Captain put another hole in projected revenue. So much so, that for the first time Odegard went to his department heads and told them to cut their costs by five per cent. Either that, or find new revenue in outside funds.

Yet even against such dire times, Odegard risked the juggling of an especially large and expensive ball whose erratic trajectory seemed certain to bring it crashing to earth, taking a good piece of UND Aerospace with it.

"Someplace along the line," says Tony Grainger, "John decided we needed a super computer. That was really high tech, he'd say. It was going to put us on the map."

In fact, ever since the federally-funded building known then as CAS IV had been in the planning stages—ostensibly to house the Earth Systems Science Institute—Odegard had mentally reserved a large chunk of its space for a so-called super computer. The term "super" referred to the number of processing chips inside the machine and the speed at which it could process information—something on the order of 500 million instructions per second. But it was also a super-large, room-sized machine that required its own room-sized air conditioning system. It was also super-expensive, in the $10 to $12 million range.

The biggest name in super computers at the time was Cray. Odegard tried unsuccessfully to get the company to donate one. He then went looking for a "previously owned" super computer. "He wasn't afraid to ask for anything," says Jerry Murray. "He spread word around the industry that he was looking for a super computer because UND Aerospace was developing new programs and initiatives that needed a super computer to make it all work."

In an interview with the Alumni Review magazine, Odegard said, "It is almost impossible to contemplate the Earth Systems Science Institute without also including supercomputing capabilities as part of the support systems."

But few among his faculty thought there was any need at all for a super computer. Even Odegard had no idea what it would be used for; he simply wanted one, says Tony Grainger.

"People were just going to come flocking to us to buy time on it," he says. "I kept trying to tell him, 'John if you don't have a use for it, the last thing in the world you want to do is invest money in computer machinery. It's going to be obsolete, and there's nothing more worthless than an obsolete super computer.'"

But such arguments didn't have any effect.

"John was tenacious," says Grainger. "He would decide this was what he wanted, and the guy would work and work and work until he got it. He just plain wouldn't let go."

He was so sure he would get a super computer that he directed Grand Forks architect Bill Schoen and his partner Jim Kobetsky to design the CAS IV building with a super computer in mind. That meant not just designing a room large enough for the computer, but another room for the air conditioning system. One of those old super computers ran hot enough not just to require its own air conditioning system, but to warm the building in the winter with its excess heat. Thus, the cooling and heating systems for the entire building were designed around a super computer that the school didn't even have.

"This was a piece of equipment he wanted to show off," says Schoen. "Why shove it in a closet? The room we designed was quite open, which gave it the flexibility for tours."

All he had to do now was find someone who'd donate their super computer to the cause. One night, working late in his office at CAS, Odegard took a call from the president of Shell Oil Co., in Houston. He told Odegard he'd heard through the grapevine that someone in North Dakota wanted a super computer. Odegard went into his high-speed selling mode. By the end of the call, Shell had offered to donate to UND its 3-year-old, $11 million Cray super computer which it was about to replace.

"Odegard had landed the big fish," says Jerry Murray. "We all shook our heads and said, 'Damn, if John dreams it...'"

Naturally, that's not the end of the story.

The gift of the computer played large in Grand Forks. The *Herald* boasted that "UND has become a supercomputing power." As with the storied altitude chamber, the super computer was the only one between Minneapolis and Seattle. It was expected to spark development of a technology center adjacent to CAS. It was going to act as a magnet, drawing industries to Grand Forks, creating jobs and turning the ESSI into an international wonder.

Terri Clark, Odegard's director of fiscal affairs, remembers the excitement, and Odegard's familiar promise.

"I can still hear John saying, 'We're going to get this Cray super computer. Shell oil is giving it to us, and it's not going to cost us one penny.'"

But almost immediately there was a large expense. In a bit of altitude-chamber déjà vu, someone had to pay to get the computer to Grand Forks and get it installed and running. Odegard asked the state of North Dakota's Department of Economic Development and the city of Grand Forks to contribute $275,000 each toward that end. Both eventually agreed, but only if they could receive, in exchange, $275,000 in time on the super computer. No problem, said Odegard who planned to charge $400 an hour for its use. Richard Olson, president of the Grand Forks regional economic development corporation told the *Grand Forks Herald*, "I don't know of any place in the country that wouldn't be thrilled to get a super computer."

As the Cray was being installed in the nearly-completed CAS IV, the state Board of Higher Education announced it would officially name the building after Tom Clifford. The man who'd been president for 21 years—and a mentor to Odegard even longer than that—was about to retire. But before he left office, Clifford left one last reminder of his feelings about aerospace and his protégé. During a graduation ceremony in June for China Airlines students, Clifford presented Odegard the President's Medal for "his service to UND, the state and to humanity." The award, along with the arrival of the much heralded super computer, didn't reduce the jealousy or cross-campus sniping one bit.

"John had the good fortune of having the president on his side," says Kent Alm. "He wasn't going to get into trouble anywhere because Clifford liked him. You just couldn't have another president like Clifford who could get away with that. He ran things the way he wanted to."

But while Clifford planned to remain as chair of the Aerospace Foundation, he would no longer be able to run interference with the faculty on Odegard's behalf. The man who succeeded him as president, Kendall Baker, never did quite warm up to Odegard. "To expect him to understand a strange operation like CAS, takes some doing," says Alm, "especially when the whole campus says they shouldn't be there."

Meanwhile, the saga of the cost-free Cray computer continued. Once Clifford Hall had opened, and the Cray was installed, it turned out that the computer came with an expensive requirement: a full-time, on-site factory support staff at a cost of $360,000 a year. Unfazed, Odegard said he would simply pay the tab out of those $400 hourly usage fees.

The super computer was installed on the first floor of Clifford Hall, adjacent to a two story wall of glass. Odegard had a series of flood lights rigged to illuminate the computer at night. Popular tours were set up right away for school children and visiting VIPs, with much oohing and ahhing over the computer's superness: it had a memory of 20 gigabytes—an unimaginable amount in 1992. It could also perform a half billion calculations a second.

There was really only one problem with the super computer. Very few people, agencies or organizations in Grand Forks or around the state, had a half billion calculations they needed performed in a coon's age, let alone a second.

Six months went by and hardly anyone had found a reason to use the computer. Many tried it out and found that, yes, it was doggone fast. But no one brought it steady work. Meteorologist Leon Osborne was probably the faculty member most enthusiastic about using the Cray to run an atmospheric research model he'd designed. But after several trials he concluded, says Grainger, that UND's Cray wasn't the right model for his project. He actually needed the next higher model in the series.

"It looked impressive," says Tom Clifford. "But it didn't really do much."

One thing it did do, however, was stick out like, well, a super computer in the middle of all the red spring wheat and potatoes and sugar beets that North Dakota had been known for. However, at the end of 1992, The National Weather Service cited the Cray as one of the reasons it decided to locate an $800,000 regional weather forecasting station in Grand Forks.

"Our forecasters will have much greater access to post-graduate educational opportunities and research instrumentation, such as the Cray computer," the Regional director of the NWS, told the *Grand Forks Herald*. And just as city fathers had hoped, the weather station, which would create 25 jobs, set up shop on land owned by the Aerospace Foundation. It had been set aside specifically for a technology park on the western side of Interstate 29, a stone's throw from CAS.

Even so, the Cray computer was hardly ever used.

"Here he had a super computer," says Kent Alm of his boss, "he just couldn't figure out what to do with it. It was very pricey. Every time you flipped a switch it cost money." In fact, it cost money on or off.

After a year and a half, Odegard hired a full-time professor whose job was to drum up business for the computer. There was a lot of excited talk about landing research contracts to design futuristic navigation systems for cars and for studying satellite images of the earth for NASA. Yet the flood of $400-an-hour users never happened.

"Really," says Tony Grainger, "it was probably the most underutilized super computer in the country."

The cost of operating the little-used Cray was a source of controversy on the recession-weary campus, especially with a new president at the reins. Things got worse early in 1993 when CAS announced it was $1 million in the red. Odegard put in a hiring freeze and told departments to shave almost $500,000 from their budgets.

"The deficit is mainly a result of budgeted revenues not developing as projected in several areas," Odegard told the *Fargo Forum*. Odegard became so concerned about the enrollment that he hired Jerry Murray's agency in Minneapolis to launch an advertising campaign

"I remember one time sitting in a meeting with him about the fleet," says Terri Clark. "The college had a big deficit balance. There had been a downfall in enrollment, and John didn't want us to have to sell airplanes to raise money. He thought it would look bad to the world. But we had more planes than we needed. He'd hired a marketing firm to increase enrollment, and he was so characteristically optimistic he'd keep saying, 'They're going to come.' When we'd do the budget, he'd put in another hundred students and say, 'Those marketing guys better get us another hundred students.'"

In the fall of 1993, the Baker administration, noting that the university itself faced a $5 million shortfall, proposed cutting 14 programs campus-wide. One of those on the list was the Space Studies program—at the time not just Odegard's biggest enterprise but the university's second largest graduate program.

Protests came immediately from CAS. Chuck Wood, then the chair of the space studies department, lamented that not only was the department the second largest, it was actually growing in the midst of the slowdown. Odegard noted that of the department's $500,000 budget, only $60,000 came from the state.

Decisions on program cuts were to be made at a December meeting. Before hand, Odegard fired a salvo at President Baker when he said he was sure the program would remain, for "to do otherwise would demonstrate a great lack of vision." In the meantime, CAS found a partial solution that stopped short of selling airplanes. Terri Clark and UND Vice President Lyle Beiswenger worked out a strategy to refinance the fleet from a five- to a ten-year lease. In the meantime, in updating the fleet, they saved money by selling eight at a time and replacing them with only five new aircraft.

In December, the space studies program was saved, but not before an ugly incident at a Faculty Senate hearing on the budget. A *Grand Forks Herald* report of the meeting described the manner in which an English professor distinguished between programs that were purely education and those that were simply training.

"You train a dog, you train a horse, you train a pilot...," he said.

The story then told how a space studies professor replied, "in a shaking voice that she objected to students being referred to as if they were dogs or horses being trained. 'That's beneath the dignity of this body,' she said."

The professor, according to the story, apologized. But the basic core resentment of the faculty toward aerospace was now out in the open. Forgetting that accountants and doctors and lawyers—even teachers—are also trained and presumably educated at the same time, the comment was especially bizarre in light of all of Odegard's achievements to date.

Only a few weeks later, Odegard would add another coup to his long list—this one a little more dubious than all the others, though on paper it looked marvelous. Shell Oil Co. was donating a second Cray to UND—the next model up that Leon Osborne wanted. It was twice as powerful as the first, but, as it turned out, just as unpopular.

Years later, when Bruce Smith took over as dean, he took a look at that second Cray under its spotlight in Clifford Hall and noticed something odd. It had been disconnected from its power source and had been for years.

If you're not failing every now and again,
it's a sign you're not doing anything very innovative.
Woody Allen

Chapter 24
Everything is Connected to Everything Else

Viewed from a scorekeeper's narrow focus, the super computer adventure could be characterized as a failure—and a costly one. But if so, what were the ramifications? The school didn't close, Odegard didn't get fired, the earth didn't stop turning. The reality: there were no ramifications. No one ever really knew or cared that the Cray had fizzled. From the global perspective of a satellite camera, the point seemed to be that UND had a super computer and all those other places down there didn't.

Since John Odegard never pretended to be a scientist or a computer maven, the utility of the Cray project was almost irrelevant. To him, the success of the thing was not just getting a super computer—and then, quite improbably getting another one—but being able to sound a Nordic whoop about it that could be heard across multiple time zones. It was ever thus with Odegard, from day one. He chose his lieutenants on the basis of their enthusiasm for running with his offbeat ideas and wrestling the details into workable programs that advanced the mission of the school. In that long run, his system produced many more highly visible strategic successes than small scale tactical failures.

In the short run, it was the budget deficit of the early nineties that made the luxury of the seldom-used Cray computer look bad. Odegard knew that solving the problem wasn't a matter of finding more clients for the computer, but tackling the larger question that had helped trigger the deficit: falling enrollments. Odegard dogged his staff relentlessly to find creative ways to increase the enrollment. And they did, presenting him with strategies that may have seemed as far fetched as the Cray super computer at the time, but which have paid enormous dividends through the years.

One such plan came from Kent Alm as far back as 1988, a year before the number of aviation majors peaked at 1,854. Through the efforts of Senators Quentin Burdick and Kent Conrad and then-Representative Byron Dorgan, the FAA had earmarked about $3 million in funds for program development at UND Aerospace. When Odegard seemed uncertain what to spend it on, Alm suggested the development of a distance learning program

Initially, Odegard and Bob Shumaker, his associate dean, were cool to the idea. Distance learning programs had been in place across the world since the days of radio and television. But only with the development of advanced teleconferencing methods and computer networking in the eighties did distance learning begin making it possible for teachers to communicate with individual students thousands of miles away.

Alm drew up a proposal for a distance learning system at UND and forwarded it to the FAA. His plan would require an extra $4 million on top of the $3 million Congressional earmark. Alm remembers Odegard telling him he'd never get it. But distance learning was such a cutting-edge technology then that the Congressional delegation liked the idea. So did the FAA.

When Alm came back with the needed $7 million, Odegard became an instant distance learning convert. The grant was to be a joint project between UND and the Midwest Aviation Resource Consortium in Minneapolis, with technical assistance from the IBM corporation. About half of the grant was used to build a supremely sophisticated television production studio in the UND half of CAS III—the former Northwest building. It was dubbed the Aerospace Training and Research Center. The original idea was to produce training programs for Air Traffic Control students and to broadcast them live, via satellite, to FAA-approved aviation schools, a college in New Zealand and 21 U.S. Air Force bases.

The success of those broadcasts had much to do with the FAA's subsequent certification in 1992 of the CAS Air Traffic Control program. Today, through the school's award winning Aerospace Network, distance learning has expanded in several directions. One of them, the space studies graduate program, is conducted in both a resident and a distance-learning format. In fact, it is currently the largest graduate program in the entire university.

Distance learning of a different kind got underway at CAS in the late eighties when Odegard considered expanding operations off campus. When he first looked to establish an aviation program at the state university in Mayville, 60 miles south of UND, it was to relieve the congestion of all those student flights at the Grand Forks airport—brought on by the surging enrollment of the late eighties. The idea was that Mayville, where the city had just extended the runway of its municipal airport, would train perhaps 100 students in the first year of the aviation curriculum. They would use planes supplied by CAS. Mayville students would

North Dakota native and Astronaut Jim Buchli was the first person ever to receive an honorary degree from Odegard's Center for Aerospace Sciences. Assembled that day in 1992 for a memorable photo are, from top left, Tom Clifford, Jim Buchli, Earl Strinden, John Odegard, UND President Kendall Baker. Bottom left: Jean Buchli, Diane Odegard and Toby Baker.

transfer to UND in the second year and go on to complete their four-year degree in Grand Forks.

When Tom Clifford saw the potential for expanding to other small towns in North Dakota, the state Board of Higher Education gave its blessing. But when only nine Mayville students signed up for the program, it was put on hold for a year. In 1990, the program finally got going, but it was a shaky start: only 18 students. They trained on a pair of Cessna 152s flown down to Mayville in the morning and then back at night. The program was eventually dropped for lack of numbers in Mayville, but also because the sudden drop in enrollment in Grand Forks lessened the urgency of a spill-over satellite campus.

It wasn't until two years later, in 1992, that expansion began to suggest itself—not as a solution to overcrowding, but as a possible boost to a falling enrollment. At that time, the astronaut Jim Buchli had recently retired from the military and taken a job with Boeing in Huntsville, Alabama. Even though Buchli was a native of North Dakota, he and John Odegard hadn't crossed paths until the mid-eighties. By then, Buchli had made two successful flights as a mission specialist on the shuttle Challenger. On one of his visits to his home state, he and Odegard got hooked up together.

"We just met each other and enjoyed each other's company," Buchli recalls. "Ours was an evolved relationship. We spent a lot of time talking to each other on the phone. He had a very high standard in everything he did. John worried about the right things—how to get support for his programs. He didn't believe in cutting corners. His maintenance crew was second to none. There's nothing about UND Aerospace that would tell you it's a state-run school. It's extremely well done.

"I think he was 'goal-oriented' rather than 'driven.' Part of that is the upper Midwest mind set. You get folks who grew up in the 50s and 60s who were hard working and focused on being successful. Those were the virtues that were respected. My interest was 'Here's an evolving entity out in the middle of rural United States that was different. Here's a guy with enterprise; how can I help him get what he wants, and how can I contribute to this part of the United States.'"

Buchli had already shown his support for UND Aerospace by carrying the school's banner into space with him on his third and final space shuttle flight on Discovery in 1991. When he left the astronaut corps and started working in Hunstville, he learned about a year-round aviation training program there for high school students. Called "The U.S. Space Camp and Aviation Challenge," it was run by Ed Buckbee, an industry-known author, lecturer and longtime NASA hand. His program was geared toward ground school instruction, survival training and some student time on flight simulators.

"John and I were talking about it one day," says Buchli, "and we said why not have the kids really fly?"

It was an idea that had already worked in Grand Forks. Since 1983, Ken

Polovitz had run a summer aerospace camp at UND for eighth, ninth and tenth graders around the country. So Buchli and Ed Buckbee and Odegard's staff worked out the details for upgrading the Huntsville camp.

"John brought in first quality airplanes," says Buchli. "One of the best things about UND Aerospace was their great safety record. He also moved UND flight instructors down there for the summer. He had a really good concept: He brought in really young flight instructors who could relate to kids."

The program—another project of the UND Aerospace Foundation—was run out of the Huntsville airport, not far from the Marshall spaceflight center and Redstone arsenal. Each of the students got four flights during their week's stay.

"John's goal was exposure more than anything," says Buchli. "He wanted to have access to those kids when they decided where they wanted to go in terms of careers and education. And kids did come out of that pipeline and go into his program at UND."

The success of the year-round Huntsville program, which brought in groups of students for a week at a time, encouraged Odegard to look further into the strategy of satellite operations. His thought was to expand the *ab initio* program to other locales and form an alliance with a local university in each area.

In the summer of 1993, Darrol Schroeder found out that Williams Air Force Base in Mesa, Arizona, just south of Phoenix was about to close. "Willy" as it was known to everyone in aviation, had been the prime training ground for the Air Force. Everybody wanted to be stationed at "Willy."

"I was just hair-on-fire excited when I heard they were going to turn it into a civilian airport," says Schroeder. "I told John we could go down there and just take over anything we wanted."

Schroeder was able to negotiate the use of the Williams site, while Odegard's staff worked out an articulation agreement with Phoenix-Maricopa Community College. The Arizona students would spend two years at the UND satellite program at Williams—studying the aviation curriculum and learning to fly—and then transfer to UND in Grand Forks to finish their degree.

Using that model, UND Aerospace found several other takers. That same summer, CAS developed an aviation training program in Hawaii. The two-plus-two transfer articulation that worked in Arizona was set up with Honolulu Community College.

Not long after that, the University of Minnesota in Crookston, 20 miles east of Grand Forks, approached UND about jump-starting its stalled program in agricultural aviation—essentially training pilots in the techniques of crop spraying. Though its program was older than UND's, its enrollment had dropped because potential students didn't think a college degree was necessary to spray crops. The program hadn't grown, and its small fleet of airplanes was aging.

"We went there and updated things," says Don Dubuque, who today runs all of the satellite programs out of UND Aerospace. "Though there's not a big market for 4-year agricultural aviation, there is a niche market. It serves law enforcement flying and the Department of Natural Resources wildlife survey. When we started a DNR program at Crookston, it doubled their enrollment. We also exported five new planes out there equipped with Global Positioning Satellite systems."

Since those early days of expansion, UND Aerospace now operates satellite Flight Training Centers through Robeson Community College in Lumberton, North Carolina; Spokane Falls Community College in Spokane, Washington; Williston State College in Williston, North Dakota, as well as those in Phoenix; Crookston, and Honolulu. It also provides factory training for Cirrus Aircraft in Duluth, Minnesota. Each of the satellites creates a number of jobs for UND students. In Duluth, for example, at the Cirrus factory, there are 16 full-time flight instructors, most of them Odegard aerospace graduates.

"They all provide marketing for us," says Dubuque. "We're well known in the Midwest, but sometimes in other areas we are not. These programs provide an opportunity for students to stay in their local area for two years and then transfer here. We encourage them not to start unless they figure on coming to North Dakota for their final degree."

While the satellite campuses lined up future students and future revenue, what Odegard really needed in those deficit years was funds for the here and now. As usual he was making high level industry contacts at his normal mach-schmooze speed. But while Odegard thought of the industry mostly in terms of aviation, he was still the dean of a large and impressive group of atmospheric scientists. Whether or not scientists would appreciate their discipline being called an industry, it is certainly a field that depends heavily on its connectability to sources of federal funds. Odegard knew he needed someone as well connected as he was in Washington from an academic standpoint.

Which is why, in 1993, George Seielstad's application to become Odegard's assistant dean of academic affairs fairly jumped to the top of the resume pile. Seielstad was a Dartmouth graduate who had earned his Ph.D. in radio astronomy from Cal Tech. After 19 years there he was named director of the National Radio Astronomy Observatory in West Virginia. Seielstad belonged to that rarefied world of astronomers known as cosmologists. They study the detailed structure of quasars and other distant objects—although quasars themselves are so distant from earth that the light astronomers detect from them today started on its journey to their telescopes billions of years ago.

When Seielstad arrived in 1993, the space studies department was trying to get the Earth Systems Science project going—it was, after all, the academic rationale that got CAS IV—now Clifford Hall—built. Seielstad put together a program and when Odegard promoted him to associate dean he also named him the director of the ESSI.

A few months later, in 1994, Dan Golden, then head of NASA, came to Grand

Forks at the behest of Sen. Byron Dorgan, who'd bragged for some time about what had sprouted on the prairie. Seielstad knew from his own experience that just getting a big shot like Golden to come for a visit was practically a guarantee of future rewards.

"The advantage of being here is the image people start with of North Dakota," he says. "Their expectations are very low. Then they see the facilities and they think, 'Wow, this is as good as anyplace.' I've seen that happen a lot. It's a great weapon to get key people to visit. They go away saying, 'I had no idea you had things like this.'"

Thus it was with the head of NASA who saw, among other things, the legendary Cray super computer. "He was one who was overwhelmed by what he saw here," says Seielstad. "Just blown away."

At a subsequent luncheon, Golden sat at table with Seielstad, Odegard and Dorgan. Golden noted that NASA currently didn't do much at all in the region. That was Seielstad's opening. He launched into a discussion of ESSI and how he thought it could align perfectly with NASA's recently announced research project: "Mission to Planet Earth."

That program had been designed to study all of the knowledge that had been gleaned from space travel about earth. Seielstad proposed doing more than research and study, but wringing from the available data practical applications that had an immediate benefit to society. His idea sprang from a life of looking at the universe through a telescope and then seeing pictures of the planet earth as taken by astronauts.

"When you first see the pictures of when the astronauts went to the moon—that blue moon—all of a sudden you realize it's all very finite and everything is connected to everything else," says Seielstad. "It's a system. But in academia, we've always treated it as a biology discipline or geology or oceanography or meteorology or a polar ice study. Yet they all function together. We need to try to understand it as a system."

This is what Dan Golden heard at lunch. He was impressed enough that within two months $350,000 was forthcoming to UND from NASA—millions more would follow over the years. Seielstad used the funds to put together an alliance called the Upper Midwest Aerospace Consortium consisting of scholars from eight different schools in five states: Idaho, North and South Dakota, Wyoming and Montana.

Do not spin this aircraft.
If the aircraft does enter a spin
it will return to earth without further attention
on the part of the aeronaut.

From the first handbook issued
with the Curtis-Wright flyer

Chapter 25

You're Not Going to Forget That Flight

On Groundhog's Day in 1989, the *Grand Forks Herald* used an indelicate metaphor to describe the recent growth at the Center for Aerospace Sciences. It characterized the flurry of construction and equipment purchases—all aimed at meeting an unexpected wave of enrollments—as a "crash program."

The word "crash" is not a good term to use around aviation people, and it was especially inappropriate at UND whose safety record had been exemplary for years. But in fact, during a four-week period at the start of the year, UND Aerospace recorded four minor airplane mishaps involving student pilots. They were minor in the sense that no one was injured, although they resulted in about $50,000 in damage to aircraft.

Within a month-and-a-half, the total of such accidents had risen to seven. There hadn't been a fatal accident at the school since 1983—a superlative, odds-beating record. Still, every year, three or four non-injury incidents would occur, from the scraping of a wing against a hangar door—so-called 'hangar rash"—to skidding off the edge of a runway or even flipping over while landing in a stiff wind. Though no injuries resulted, seven such incidents in less than three months was the largest run of flying trouble at the school in 17 years. It added another level to the downside of rapidly increasing enrollment: not only were some students not getting to fly, but those who did were facing evermore crowded skies over Grand Forks.

That spring of 1989, students were expected to log about 60,000 hours of flying time, representing more than 200,000 takeoffs and landings. Within a year, the flight training hours were projected to jump to 100,000 with 350,000 takeoffs and landings. In 1990, the Grand Forks airport set a record of 2,211 takeoffs and landings in an 18-hour period, making it the nation's 38th busiest airport, according to the *Herald,* ahead of Orlando and Salt Lake City.

The string of accidents in 1989 prompted then-director of flight operations, Dana Siewert, to tell a reporter, "We are very close to reaching our maximum capacity." John Odegard quickly set up a committee, headed by his new associate dean, Bob Shumaker, to review all safety procedures. He ordered a three-day stand down in student training while each of the 150 flight instructors flew a check ride with a certified flight examiner. The idea, Shumaker told *The Herald,* was to make sure that "all instructors are teaching their students the same thing."

All seven accidents had occurred on the runway, telling Shumaker that students needed more training on landings. Based on Shumaker's findings, Odegard ordered that CAS flight manuals be rewritten to clarify runway procedures.

"Overall," Shumaker told reporter Steve Schmidt, "my preference would be to tighten up the ship a bit. A taut ship is a happy ship and, I think, a safer one. There's sometimes a tendency on a young or inexperienced pilot's mind that once the wheels hit the ground there's a great sigh of relief—I've cheated death again— and then forget the airplane is still under aerodynamic forces. We really want to emphasize that the airplane be flown all the way down the runway all the way to the chocks. I think we're making progress."

By a taut ship, Shumaker wanted to see more training discipline instilled on students by their flight instructors. He also wanted to see a series of required student safety lectures. Those procedures were put in place and, in the meantime, one instructor was fired and another suspended.

As he had after most accidents, Odegard displayed a mixture of anger at the failure of his system to prevent accidents, and constructive criticism trying to find a way to prevent them from happening again.

"When I was a student, we had a glider accident," says Bob Muhs. "And John was not happy with what happened in the context of procedure. But then it turned more into a correction. How do you teach students to avoid this? He was very methodical. The accident had exposed some weakness in department policy. So he would look at it from the context of: *Have we taught the things we need to teach?'* and *What do we have to change in our organization to learn?*

While the string of incidents in 1989 positively altered the flight training procedures at UND Aerospace, it took another five years before Odegard really got to the bottom of his question about organizational changes for safety's sake. In 1994, a series of commercial airline crashes with heavy loss of life stunned the industry. The U.S. Department of Transportation held a summit in Washington, inviting all major airlines to devise methods for zeroing out aviation accidents. Because of the reputation of UND Aerospace, Odegard was among those invited.

He spent three days in Washington and learned that the aviation industry and commercial airlines were going to mandate the appointment of a person within their organizations whose responsibility was aviation safety.

When he returned to Grand Forks, Odegard summoned Dana Siewert to his office and told him, "If the aviation industry and the FAA are going to require that, then we're going to require that here."

Siewert had been the director of flight operations since 1985. He was one of the young lions Odegard had brought along from the very early years—along with Al Palmer and Don Dubuque—pushing each to finish his college education and rewarding them with responsibility.

"We need to find somebody to be our safety director," Odegard told Siewert.

"Do you have anyone in mind for that?"

"Yes," said Odegard, "yes I do. You."

Siewert was at a loss. He wondered what he could possibly do that wasn't already being done.

"What do I know?" he asked, protesting that all he knew was what he had learned on the job.

"You probably know more than most of us know," said Odegard.

It was the ultimate compliment. Likely, Odegard also was speaking from what he had learned on the job—learned from Siewert himself.

In 1974, when Siewert was hired by Odegard as a flight instructor, he remembers being intimidated by his new boss. "I was just in awe of his presence, this gentleman who was chair of the department. He was over 6 feet tall, a big guy. Very well dressed, very well spoken. Here I was this young guy, just starting in."

In addition to flight instructing at UND, Siewert took a part-time job with the local Grand Forks FBO, pumping gas and washing planes. Not long after, Odegard called him into his office and told him he was doing a good job. He said, however, he wasn't going to go anywhere in an educational environment without a degree.

"He kept pushing and pushing," says Siewert. "I started to go to classes at UND as a part-time student, and I'd work full time. I'd knock off a class in the fall and spring and summer. It took me forever, but I finally graduated in 1984, ten years later."

During those years, Odegard found opportunities to mentor Siewert. When Tom Ryan donated the second Citation jet to the program, Odegard had Siewert fly in the right seat with him on many occasions. It was a huge deal.

"When you start to fly a plane that is burning jet fuel and doesn't have propellers, you've sort of reached the top of it," says Siewert.

The steady co-pilot assignment gave him not only the chance to qualify on a jet, but to observe his boss up close.

"You had to get in rhythm with John," he recalls. "You had to actually get John to slow down a bit. John's mind was way out in front. He was always thinking

Terri Clark (left) speaking with former UND president Tom Clifford, became Odegard's no-nonsense director of fiscal affairs in the late 80's. Equally adept at juggling finances and personalities, Clark continues today with a parallel role at the UND Aerospace Foundation.

Dana Siewert (left) Don Dubuque (center) and Al Palmer all started out with Odegard at roughly the same time in the early 70s. They rose through the ranks and today represent the top leadership in The Odegard School's airport operations.

ahead. There are specific things you have to do as co-pilot and initially it was a challenge. Getting into that right seat with him, you were already behind.

"It didn't take me long to realize I needed to bring a yellow pad along, because he'd be saying we need to do this and we need to do that, and get that done and call this person. He was always thinking about stuff. We'd fly to L.A. to pick up some China Airlines people, and I'd end up with six pages of stuff in my pad. I'd get back and go through it and end up calling someone in, say, space studies. I'd say, 'I know this has nothing to do with me, but I'm just passing on something John said that relates to space studies…'"

On one of their early Citation flights together, Odegard and Siewert flew to Roseau, Minnesota to pick up a premature baby and a nurse. They flew them to a tertiary care hospital in St. Paul, landing at the downtown airport at 2:30 a.m.

"It was just the two of us flying back," Siewert recalls. "John was in such a hurry as we taxied out. I'm trying to catch up on the flight checklist to make sure we'd done everything. We taxied out, and I remember lifting off and looking out. There was no runway left. And I remember thinking, 'Wow that was a really short runway.' And even John said, 'Wow that runway was really short.' I looked at the heading indicator and realized we'd taken off on the wrong runway."

Siewert initially found it hard to question Odegard's decisions.

"As I got to fly more with him, I got to where I could say, 'John let's slow down here, we're not ready to go yet.' It was getting up the courage. John had a lot of stature. You just had the impression you're not going to question John Odegard. But he'd want to take the most direct route to someplace, and you'd say, 'Well John, the weather is bad that way. We can't do that.' He'd say, 'I don't want to go miles out of my way.' I'd say, 'John, that's the way we've got to go.' And he'd say, 'Okay, you're right.' I thought that was good. That brought our relationship a lot closer. You could tell John something, and he would listen. He was respecting my decisions."

Thus did Odegard convince Siewert he was the right man to be safety director. With many misgivings, Siewert enrolled in a safety curriculum at the University of Southern California. He took courses in aviation safety management and accident investigation, supplementing them with courses put on by the Airline Pilot's Association and one at the National Transportation Safety Board on fatigue in aviation. When he came back to Grand Forks, Siewert was overflowing with ideas, and he immediately began putting together a safety program, starting with a list of new priorities.

1. The safety director cannot wear two hats.
Following the DOT safety mandate at that Washington summit, many airlines and aviation organizations made the job of safety director an extra duty

assignment for a pilot or a manager. Siewert realized he could not be both director of flight operations and director of safety—not because of any workload burden but because of the inherent conflict of interest.

"The duty of Flight Ops is production," he says, "to maintain planes and fly them. Safety's function is protection. You've got production versus protection. You can't put that burden on one person. You've got to find a balance of production and protection. Too much production, you take risks and subject people to having accidents. If you're too safety conscious, you don't have enough production, and you don't meet your goals, and you have financial problems."

That's why Siewert's office today is directly across the hall from the man who succeeded him as director of flight operations, Al Palmer. Neither job answers to the other; both report directly to the dean.

2. To create a safe environment, you need to develop a safety culture everyone embraces as their own.

"In a culture of safety," says Siewert, "I don't own the safety program. It belongs to everyone here. Everyone is responsible for safety not just me. That is the most difficult thing to get across. The old school of thought said that if you had a safety problem, the problem resides in the safety office. In reality the problem may lie in the organization.

"When you start showing that even though pilots make mistakes, it may be because there are certain things that are not happening in the organization, you then get people to start buying into the idea that safety is everyone's responsibility."

3. Safety procedures need to be written down

"Before 1995," says Siewert, "we thought we had a safety program. When I go to various universities now, they want to embrace safety, to have someone come in and tell them if they're doing something wrong. The first thing I always ask: Do you have an aviation safety program? The answer is always, 'Yes we do.' I say, 'I'd like to get a copy of that.' And they say, 'A copy of what? I say, 'Of your safety program.' They say, 'Well we don't really have any copies.' I say, 'Well you just told me you had a program. What is the program?' They say, 'Well, we expect everybody to be safe. We hang some signs up…'

"My message is that you have to have certain expectations, and everyone has to be on the same sheet of music. You've got to develop a written program shared with everyone in the organization—from the dean to every student—so they understand what the safety culture is, what is expected. There have to be specific polices and procedures and it has to be written. That's what we've developed here, a very specific aviation safety program. It's in writing. I can give you a copy. Is it putting safety signs on walls? Yeah, that's part of it, but not all of it. If you have an accident, hanging signs isn't going to change people's attitudes."

4. You need a safety director to distinguish between an idealized vs. an actual safety culture.

"Idealized is what we say we do," says Siewert. "Don't let people fly when fatigued. Question: is that what actually happens? You've got to take a look at what really happens in your program and compare that to the idealized version. You have to realize that humans naturally make mistakes. It's normal, and it's the same in aviation. That's why it requires a safety department, a person who can take a look and see what's really happening. You've got to have someone reminding everyone that it's not how many hours we fly, but how many hours we fly safely."

Changing attitudes, he says, is hard work. It means having good programs, getting people to buy into them and recognize them when they do a good job.

"It's a real challenge to a younger age group," he says. "They believe they are invulnerable."

5. Not having any accidents doesn't mean you have a safe program. Nor does having an accident mean you're unsafe.

"Car accidents aren't real big media items," notes Siewert, "because they happen so frequently. There's probably a fender bender that took place someplace today in Grand Forks. We're probably not going to hear about it because it's not newsworthy. But if we run an airplane off the runway and no one even gets hurt, that will be in the newspaper. Accidents in aviation are very infrequent, so when they do happen, they become a big news item.

"Ultimately, I don't think there are any aviation experts. I am not a safety expert, and I don't know any. If we were experts, we'd figure out how not to have any accidents. No one has figured that out. I don't know if we'll ever figure it out as long as we have the human element in the cockpit. Just like you'll never stop auto accidents."

The year Siewert developed his program, the FAA presented UND with its annual safety award for an accident- and incident-free year. The superintendent of Fargo's regional FAA office praised Siewert's work and commended UND Aerospace for a remarkable record, saying, "We haven't had this kind of experience with any other institution."

Sometimes, though, even a taut, safety-conscious organization runs up against Murphy's Law. Such was the unnerving case during that summer of 1995, when John Odegard, as he had done for several years, invited several friends and VIPS to fly with him to a hunting and fishing lodge in Canada. His favorite spot was Oak Lake, but the ownership of the lodge there had changed. Odegard was trying a new spot at Snow Lake, a small mining town in the Northwestern region of Manitoba. The tourist information tells you it's between Thompson, Flin Flon and a place called The Pas. In other words, it's very far away—four-and-a-half hours by air, north of Winnipeg.

The only way in and out was by float plane. By 1995, Odegard's fleet included several planes configured for both land and water landings. It's hard to tell which came first, UND owning several float planes, or the idea that flying up to a roadless Canadian lake retreat to go fishing would require one. However, it's been said many times about Odegard that if there was a plane he wanted, he could justify it. At any rate, his annual VIP fishing trips were well known in the aviation world.

Odegard started these summer escapes as a way to schmooze key supporters of the program from the Grand Forks area. He gradually expanded to include CEOs of aviation companies, airline executives and anyone in the country he thought he could do business with. Once up there among the trees and the bees, Odegard had a captive audience.

"John realized very early that it doesn't make any difference what business you're in, relationships are important," says Russ Meyer, the former chairman of Cessna. "He did everything right, and he didn't leaved any stone unturned."

As it turned out, Morocco's General Kabaj was particularly fond of the Canadian walleye. John McCain liked his trip so much he came back for a second.

"He'd swing a lot of deals that way," says Kent Lovelace.

And he wasn't shy about it. "If someone was influential," says Bob Buley, a frequent fishing trip invitee, "John wasn't afraid to ask."

That summer of 1995, Odegard flew to Snow Lake with Bob Buley, Jim Buchli and several guests from Honeywell. They flew up in a Piper Navajo. The mood was light and relaxed. Lots of hijinks. Just before boarding, Odegard loudly told Jerry Murray to take the right seat because, "I don't want some bleeping astronaut telling me how to fly my plane." During the flight, the astronaut hollered out from the back, "Where the hell are we anyway?" Odegard said he'd have his co-pilot get a fix. As Buley was studying the charts, Buchli shot back, "Oh that's all right; I can't tell where we are unless I can see the curvature of the earth."

Once on the ground at Snow Lake, however, the mood became a little tense. The group had been told by the resort people that fuel would be available there for their plane. "But we found out there was no fuel," says Buley. Meaning no fuel anywhere. It not only killed any chance of quick day-hops to other lakes—a favorite activity of Odegard on vacation—it meant they might not have enough fuel to make it back to Winnipeg.

"We had to then calculate," says Buley. "Can we make it back to Winnipeg with our reserves? Or should we fly east, over to Red Lake, Ontario—which was completely off course. We looked at the situation, and we looked at the winds. It was the most demanding flight we could have had. We had to manage fuel like we were going across the Atlantic. We had calculated our fuel from the time we took off to the point of no return. Once we got past that we had no alternative. There were just no airports to bail out to, so we'd have to put it down on the tundra.

"We had to burn every drop out of the tanks. We would stay with a tank until John started to see fluctuation in the manifold pressure. My job was to keep both hands on the booster pumps. As soon as we heard the engines starting to suffer fuel exhaustion, I'd hit the booster pumps and John would switch tanks. We did that three times. We made it to Winnipeg, but it was tense. You aren't going to forget that flight."

In Buley's case, there is one more reason for that trip to stay in his memory. It would be the last time he ever flew with his good friend Odie.

A week later, Odegard was back in Canada, this time for a staff retreat in Minaki, Ontario, just north of Lake of the Woods. Like the Oak Lake fishing trips, the staff retreats were an annual affair, a chance for Odegard's department heads and program directors to bond. Usually 20 to 25 people would fly up in one of several UND planes. These trips, however, were not all fun and games.

"There was always an agenda," says Terri Clark. "You had to prepare for it ahead of time. Usually some sort of presentation was required."

Not that there wasn't time for fun. In fact, fun was required.

"We had to play golf," says Clark. "The first time we went, I tried hard to beg my way out. Nope. I had to do this. My own assessment of why John made us do it—it was humbling to everybody. It was certainly humbling for me. None of us golfed enough to be good at it—we worked way too many hours; John told us once if we weren't working 55 hours a week, we weren't earning our salaries."

Actually Odegard wasn't a golfer either—he never had the time—and so he duffed around Minaki like everyone else. That day, Dana Siewert was teamed up with Odegard as a golf partner.

"I could tell there was something on his mind," he remembers. "He just wasn't as jovial. He wasn't having fun."

That night at a mock banquet Odegard handed out gag golf trophies. After the dinner, he played cards into the night. That retreat was the first one Dr. Warren Jensen had attended. In fact, Odegard had asked him to run it. Jensen had a room in a two-bedroom suite right next to Odegard's. Jensen thought it curious when they'd arrived in Minaki, that Odegard told him to "come stay near me."

Jensen had gone to bed before the card games broke up. At about 1 a.m. the next morning, he was awakened by a knocking on his door. It was Odegard. He said he had just passed blood in his urine. He hadn't any pain, but he was concerned.

Jensen was stunned.

In the summer of 1995, Diane and John Odegard toured Montreal with Gary Kiteley (left). Odegard and Kiteley streamlined and modernized the organizations that govern college aviation education. Shortly after this picture was taken, Odegard found out he had cancer.

John Odegard, in a rare moment of repose at the Oak Lake fishing retreat in northern Canada. He and others from the school would fly UND float planes to the lake each summer for walleye and bonding with invited aviation industry leaders .

"My first thought was renal cell cancer," he recalls. "If you have no pain, and there's blood in your urine, that's renal cell cancer until you prove that it isn't."

The next morning, Odegard complained of back pain. He thought he'd twisted something while playing golf. Though a full day's schedule of meetings remained on the agenda, Jensen told him they were leaving for home immediately. Odegard himself flew the Piper Navajo into Grand Forks. It was to be the last flight he ever made as a pilot.

PART FOUR

THE SUNLIT SILENCE

...wheeled and soared and swung

High in the sunlit silence. Hov'ring there,

I've chased the shouting wind along, and flung

My eager craft through footless halls of air.

John Gillespie Magee, Jr.
"High Flight"

Anyone can do the job
when things are going right.
In this business we play for keeps.
Ernest K. Gann

Chapter 26
The Laws of Gravity

Whatever form it takes, cancer specialists describe the disease in terms of its stage of advance. One can almost guess, by the weighty use of Roman Numerals, that a deadly illness such as cancer doesn't have too many stages. A doctor always hopes to find a tumor in Stage I. The tumor is small and confined to the spot where it began. It is usually easier to treat, with highly favorable odds for a successful outcome. Stages II and III measure the slow and then more ominous spread of the cancer to neighboring tissue and organs. Finally, deadly Stage IV. It marks the extensive reach of the cancer throughout the body's organs. The chances for survival are lower, as are the number of weeks or months one is expected to remain alive.

The cancer that Dr. Warren Jensen had conditionally diagnosed to himself after hearing John Odegard's symptoms—and that he later kidded himself might just be kidney stones—was unconditionally real.

When he got back to Grand Forks, Odegard met with his physician and long time friend Jim Brosseau. After a CAT scan showed a mass in Odegard's kidney, Brosseau contacted a classmate of his from the UND School of Medicine. Victor Corbett, M.D., was an endocrinologist at United Hospital in St. Paul who had grown up in Minot, North Dakota and knew Odegard well. While Al Palmer flew the Odegards to the Twin Cities, Corbett paved the way for treatment at United. The corrupted kidney was removed after doctors there confirmed the diagnosis that Jensen and Brosseau had feared: Stage IV Renal Cell Carcinoma. In 1995, the prognosis was a life expectancy of four months to 11 months. Three days later, Odegard was told the cancer had spread to his brain.

The next stages that John and Diane Odegard lived through—as friends and relations of any cancer victim well know—were the all too common path of hopes raised, hopes dashed, hopes raised again and dashed again and so on until one or the other wins out.

In Odegard's competitive mind, his cancer became "The Dragon" and he the defiant warrior who would slay the beast. Indeed, he raged and thrashed mightily—vintage Odegard—he caught a glimpse or two of light, but ultimately succumbed. While a hardened competitor might even say he lost, the more enlightened would, no doubt, suggest—as had Paul of Tarsus—that he "fought the good fight, finished the race, kept the faith." And if there be points given for moral victories, Odegard should be allotted his share. For the cancer that killed him took almost three-and-a-half years, an almost unprecedented survival span for Stage IV renal cell cancer. During that time he set the ultimate good example. He remained publicly upbeat, continued to work hard and enthusiastically and moved as fast as he could for as long as he could.

Stories of beating cancer are almost always uplifting. Those that describe the alternative are almost always difficult to digest. A moral victory over cancer may not mean much except to those left behind who may draw from it the strength to carry on their own good fight over their own dragons, whatever they might be.

It was no small feat that the empire John Odegard built and ran like a stagecoach driver eluding outlaws, drew from its stockpile of Odegardian strength to remain alive and healthy long after his death. Had it really been the two-dimensional, low-brow, dog-training outfit its critics made it out to be, the school probably would have failed. When the lion tamer is gone, who tames the lions? But just as Odegard exemplified E.M. Forster's round character to the end, so does his School for Aerospace Sciences in the here and now.

The school went through its ups and downs in the next three years, though the good vastly outweighed the bad. Mistakes made were mistakes caught and rectified. The founder's insistence that those around him display the same zeal and desire to help students as he did, had taken hold long before the reins slipped from his fingers. His loyalists weren't about to let his passing divert their attention from their mission.

A good bit of that carry-on spirit was imparted by Odegard in poignant moments during those final three-and-a-half years. And though he didn't go from taskmaster to easy-going guy overnight, Odegard's personality did undergo a change, almost as if he'd begun to recognize some excesses and wanted to take some steps to mend some fences.

While he'd always been a church goer—raised a Lutheran, he'd converted to Diane's Episcopalian faith—Odegard developed an open spirituality in those final years that few had ever seen before. It seemed to surface on a night in October of 1995, when John and Diane attended a birthday party for a close friend.

By that October, the Odegards had already been thru the cancer wringer. Following the July removal of his kidney in Minneapolis, Odegard consulted with his college pal Lloyd "Corky" Everson, now one of the country's top cancer specialists. Everson, who lived just north of Houston, was president of US Oncology, a company that sets up community cancer centers across the country. He told Odegard that while he could be treated satisfactorily at almost any cancer center in the country, there were two places—M.D. Anderson in Houston and Sloan Kettering in New York—that saw the most cancer cases, and thus had the most experience. He offered to pave the way at Anderson, and Odegard agreed.

Everson arranged for the Odegards to come to Houston for a preliminary meeting later that July with Anderson doctors. Then, quickly back at home in Grand Forks, they attended the August wedding of John Jr. Because of the cancer, the wedding had been moved up from its original September date. Immediately after, the Odegards returned to Houston where doctors removed the tumor from John's brain.

That surgery was so successful that the Odegards were back in Grand Forks only a few days later. They were in time for the August CAS graduation ceremonies, where Odegard personally handed out the diplomas. Students and faculty alike were shocked and moved to see him there so soon after brain surgery. As a precaution, Dr. Warren Jensen stood only a few feet away in case Odegard fell ill, but the event went smoothly.

Weighing on Odegard's mind was a decision he'd made in Houston. He and Diane had stayed at the home of Corky and Jacque Everson and spent hours discussing whether or not John should fight the cancer.

"He was trying to make up his mind about what to do," recalls Everson. "I told him that what it required was an aggressive, highly experimental approach. I said no one knows if that will work. You may go through it, have all the side effects, and it may not do anything."

Warren Jensen, who was present as well, remembers how Everson helped Odegard understand the psychological issues he'd be facing.

"Corky would say, 'We can do chemo. We can do a lot of things,'" recalls Jensen. "'The question isn't always what can we do, it's what should we do. What is the quality of life?' He said that if John went the chemotherapy route he should not fight it, but psychologically accept it and take it into his body. He said, 'If you're going to do the chemo, accept that it's your choice. You can't sit and say the damn chemo is making me sick. The chemo is doing its job. That's what it's supposed to do.' Corky was very good that way."

After several emotional days of weighing his options, Odegard reached a decision.

"Finally," says Everson, "in typical fashion, he said, 'I'm going for it. I want to do everything I possibly can. I'm not giving up.'"

It didn't surprise Jensen.

"John didn't feel the odds applied to him," he says. "Kind of like the laws of gravity didn't apply to him because he could fly. He also didn't think you could take group statistics and apply them to an individual. An 11 month average survival meant to him that some people lived longer. And some beat it. So he wanted to know how he could optimize his ability to beat it."

Following the CAS graduation in Grand Forks, the Odegards returned to Houston for the first phase of John's chemotherapy. The staff and faculty at CAS had prepared a surprise sendoff for them, waiting en masse at the airport with cheery signs wishing him well on his trip.

Just as he was leaving his office on campus for the drive to the airport, he got a call from a sobbing, grief-stricken Xavier Ziegler. He was the executive with Airbus who had hosted the Odegards at his villa on Corsica. Ziegler was in Minneapolis, calling because he had just learned that his son had been killed in a climbing accident on Corsica. Bob Buley was also in Minneapolis that day, at work at Northwest. He remembers Odegard's frantic call to him moments later.

"He's not quite in tears, but close," Buley remembers. "He says, 'Bobby, Bobby, Bobby, something awful has happened. Xavier's son was killed in Corsica. Bobby, we've got to take care of Xavier.'"

Buley helped arrange passes on Northwest to get Zieglar back to Europe, but can't forget Odegard's concern. "It would have been understandable if John had been totally focused on John," he says. "But he kept bringing up the tragedy of Xavier's son."

When they got to Houston, the routine surrounding the chemotherapy treatment became unnerving, recalls Diane. The skin fell off the soles of John's feet, and his skin itched terribly. He countered it by regular workouts and endless walks with Diane around the hospital grounds. As a change of pace, they would go on outings with either the Eversons or Jim Buchli and his wife Jean.

The plan Odegard's doctors had worked out called for him to go back to Grand Forks, continue taking the chemo locally, and return to Houston every three months for a checkup. Thus, he and Diane returned to Grand Forks the first time during the fall 1995 Homecoming celebration at UND.

"It was the biggest and most exciting gathering for the Center for Aerospace Sciences," says Diane. Along with Stephanie and John Jr., close friends Jim Bunke and Jerry Murray were visiting as they had done at several Homecomings past.

"John was tired and even dozed off at a point," says Diane. "The evening loomed before us. What should we do? Should we go to the banquet and make an appearance? This seemed impossible in my opinion. Stephanie wasn't so sure either. But it was an important evening for the aerospace family. It had always been the most upbeat, happy time for us."

Bunke and Murray left for the banquet and John Jr. disappeared upstairs with his father.

"Shortly thereafter," recalls Diane, "John Jr. called down and said to get ready for the banquet. Then John and his father appeared on the stairs. Odegard was fully dressed and as dapper looking as usual—thanks to a hand from his son.

"It was a moment I'll never forget," says Diane. "It was the first time he had gotten dressed for a formal occasion since the earth caved in on us."

With their grown children by their sides, they left for the Ramada Inn.

"The moment we entered the room," Diane remembers, "we were surrounded by well wishers, people truly happy, but somewhat shocked to see John for the first time since the illness and surgeries. We weren't alone for a minute."

They left for home shortly afterward. "We couldn't stay very long," she says, "and when we drove away, I knew things had changed forever. I looked at John and said, 'That was one of the hardest things I have ever done.' We were both near tears as we drove home in the rain."

Only a few days later, the Odegards made a second social foray, turning out for the 50th birthday party for Ken Svedjan's wife, Loretta, at the home of good friends Rob and Judy Larson. At the party that night, a guest named Robin Silverman noticed Odegard sitting alone in a corner. She'd had a nodding acquaintance with the Odegards, seeing them at various community functions. She'd always found John to be intimidating. "He was very large and in charge and always had the grandest stature in any room," she remembers. "I had never approached him before at any other function."

But she was intrigued by what was being said about his cancer—that the odds of beating it were a million to one. Silverman had been a local newspaper columnist and had written several books on the power of the human spirit and the metaphysical power of the mind.

"He was sitting in a corner on the porch by himself," she recalls. "I went over to him and said, 'What if you're the one in a million?' He goes, 'What?' I said, 'There's always someone who wins the lottery. Someone who gets well from cancer. It could be you.' He says, 'No, I'm done.'"

Silverman, who is neither doctor nor nurse, then began telling Odegard about medical studies at several major universities on prayer and healing.

"According to doctors, you are done,'" she said, but told him about studies that showed patients how to get positive help from prayer. "I talked about visualization and the mind-body connection. The mind is a very powerful tool. He clicked into it right away. Those blue eyes sparkled. He said, 'Tell me more.'"

In November, the day the Odegards flew back to Houston for their first checkup, Robin Silverman got a call from Loretta Svedjan. Diane had been visiting with her and asked if their friends would pray for John.

"I said I have a better idea," recalls Silverman. "Let's get the whole town praying for him."

This she did. Silverman, who is Jewish, wrote a non-denominational prayer for her newfound Episcopalian friend and faxed it to various newsrooms in the area. She also passed it along to friends. By the time the Odegards got back to Grand Forks from that first checkup, prayer chains had sprouted across town and beyond. Several people had sent the prayer to friends across the country. Friends, relatives and those who knew Odegard only by reputation would spontaneously gather in offices and homes—even in the offices of UND president Kendall Baker—at the suggested times of 10 a.m. and 10 p.m. on given days and recite Silverman's prayer.

"It was an amazing public kind of thing," says Diane. "Everyone wanted to help."

Later, a psychologist friend of Everson's from Fargo named Dick Harper met with Odegard and expanded on the mind-body connection. He prepared audio tapes to listen to in Houston to help visualize tranquil and positive scenes during chemotherapy. "They were really helpful," says Diane. "Dick made a great series of tapes. Then Robin made tapes for many months. We listened to them every night."

The Odegards gradually settled into a routine between Grand Forks and Houston. John's brother Jim, who was then the state's attorney in Grand Forks, lived only two blocks away. His wife, Cari, was a nurse and offered to give John the daily chemotherapy shots he needed.

"She came into the house every morning even before we were up," says Diane. "She got the drug out of the fridge. She woke us up. She had a wonderful bedside manner. She sat down on the bed and gave John the shot. This made a big difference. She'd come back at night and give him other shots. She did it for all three years. She saved our lives basically."

During his various Houston checkups, doctors found evidence of cancer on Odegard's spine and in his lymph nodes. He underwent four more major surgeries—many in the 9 to 10 hour range. But by the summer of 1996, tests began to show a diminishing of the cancer. And by fall, Odegard had shown so much progress that his optimistic doctors temporarily halted the chemotherapy to give him a chance to heal from his surgeries. He went back to work, and it began to look as if he just might beat it.

"I remember he told me one day, 'I'm cancer free. I've beaten it,'" says Dana Siewert. "He said he was going to get his medical certification back so he could fly again." During a radio interview that November, Odegard sounded as upbeat as ever and told his audience, "I'm doing extremely well. I'm deeply blessed by people offering prayers for me. I'm here to say they worked. I'm looking forward to getting back to flying airplanes."

He even asked Jean Haley Harper, a part-time freelance writer and full-time United Airlines captain, to write a piece for the CAS newsletter, letting everyone know he was fine. "I'm completely cured," he told her. "I want you to write it so positive that there's no doubt in anybody's mind that I'm here to stay."

At his doctor's recommendation, Odegard reluctantly started his chemotherapy again, with the idea of making sure all of the cancer was gone.

"His psychological attitude allowed him to take on a tough chemotherapy," says Warren Jensen. "But it was very difficult physically and emotionally to go through. John's strength of character and determination helped him the most. Diane was his number one supporter."

The Odegards looked forward to spring. In April, John would go to M.D. Anderson for a final check and stop the chemo altogether. He and Diane would then take a long vacation in Arizona.

On April 4, 1997, it started to snow in Grand Forks. With winds blowing at 44 miles-an-hour, the storm was 9 miles-an-hour above official blizzard status. This was the eighth such blizzard to hit the town that winter. This one, (The *Grand Forks Herald*, which gave names to the blizzards, called it Hannah) lasted three days. It brought the level of the Red River of the North, which separated Grand Forks from East Grand Forks, Minnesota, to flood stage at 28 feet. No one panicked. The river often reached and went over flood stage. It was held back from the two cities by a series of old earthen dikes.

Two months earlier, in January, in the Regional Weather Information Center at CAS, Leon Osborne did a risk analysis on those blizzards, looking to the potential of spring flooding. His estimate was that the Red River would eventually rise to 52.5 feet, which was six feet higher than the estimate of the National Weather Service.

The NWS scoffed at Osborne's prediction, noting that the highest the river had ever crested at was 46 feet. The Army Corps of Engineers joined the criticism. According to Osborne, the NWS told the North Dakota congressional delegation to disregard his prediction, even guaranteeing that the river would not go that high. The reason the numbers were so important was that the system of dikes that ringed Grand Forks were 52 feet high. Osborne, in essence, was predicting that the river would overspill the dikes and roll into town.

A few days before blizzard Hannah hit on April fourth, John Odegard invited the U.S. Secretary of Transportation to tour the city. "John was just beaming," says Osborne. "I can still see his face when the secretary asked me how bad the blizzard was going to be, and I said it's going to shut the state down. He turned and looked at John, and John just smiled and said, 'It's going to happen.'"

The storm indeed put a stranglehold on the area for several days. After the storm passed, temperatures rose and all that snow started to melt. By Wednesday, April 16, the runoff of the melt had pushed the Red River above the old crest record of 46 feet. A flood seemed a certainty.

John Odegard's home on Reeves Drive sat two blocks from a section of the

Against the advice of his academic counselors, Leon Osborne left a
graduate program at the University of Oklahoma in 1978 to manage
weather data for Odegard. He became a major creative force in
Odegard's nascent science initiative. He is the founder of the Regional
Weather Information Center that helps both farmers and travelers.

earthen dike, clearly in harm's way. By Friday the 18th, most of the city's 52,000 residents were already being evacuated. The university had been shut down and all CAS employees were long gone. John and Diane were preparing to catch the last flight out of Grand Forks for their big—and hopefully final—checkup in Houston.

That Friday, at almost the last minute, Leon Osborne learned that if the university's Regional Weather Information Center shut down during the flood, the school would likely lose future funding from the Federal Highway Administration. That's because the center provided essential en-route traveler weather information that was a key part of a FHWA research program. (The successful program later became the blueprint for a national advanced traveler information system known as 511).

Uncertain what to do, Osborne evacuated his house and moved his family to the weather center in the CAS I building, several miles west of the river. There was no one else around to run the facility, which was supposed to provide weather information 24 hours a day. Even the National Weather Service had evacuated its facility in town, making Osborne the sole weather voice in Grand Forks.

Suddenly into the weather center came John Odegard. He wanted to make one last check before he left town. When he realized that Osborne and his family meant to stay there for the duration, he asked where they intended to sleep.

"We had some blankets," says Osborne. "John said this won't do. He then left and found a sporting goods store that was still open. He bought all of the sleeping bags they had, the high end bags, and brought them back here. He also had a full supply of bottled water with him. He did everything he could to be sure we were comfortable. Then he left. When he was in Houston, he called everyday. That meant an awful lot."

Though his wife and two younger sons were able to relocate to nearby Larimore before the river crested, Osborne and his oldest son stayed behind; the father forecasted non-stop for nine straight days, and the son provided general assistance, including a constant checking of the water level in the mechanical room of CAS.

The river spilled over the dikes that Friday—as Osborne had predicted—and didn't fully crest until Monday, at 54.11 feet. While it vindicated Osborne, it wrought such havoc and misery on the city that at one point its streets were not only covered by the raging river, but part of it was also burning to the ground. The water had shorted out electrical systems and caused a fire in the downtown commercial area. Fire trucks couldn't reach it. Some of them were mounted on flat bed trailers provided by the military and towed into place so as to aim their hoses at the unreal sight of shooting flames above a river that had turned into a sea. In all, 11 buildings were destroyed by the fire in the downtown area, including the newspaper offices of the *Grand Forks Herald.*

In Houston meanwhile, John and Diane were joined in their guest apartment

Xavier Ziegler, right, was an executive with the French airline
company Airbus when he met Odegard. His and Odegard's
families became close and vacationed together at Ziegler's home
on Corsica. A heavy smoker, Ziegler died of cancer not long
after Odegard.

After Odegard was diagnosed with cancer, his old college buddy, Lloyd "Corky" Everson,
a national cancer expert living in Houston, connected him with the M.D. Anderson clinic
for treatment. John and Diane spent much time in Houston often socializing with Everson
(far right), and Houstonite Jim Buchli and his wife Jean.

across the street from the hospital by their children, Stephanie and John Jr. Odegard was his old impatient self, unable to bear the suspense of waiting to hear his test results. By then, Odegard had been to Houston often enough that he'd managed to find out from a hospital employee that test results were dictated by lab personnel into a telephone recording system. It was accessed by doctors who then relayed the information to the patient. On a previous trip, Odegard had uncovered the secret phone number and had called it several times to get his test results before his doctor. Once, he made such a call from a restaurant. He had a hard time understanding the recording, so he passed the phone across the table to Jim Buchli to see if he could make it out.

That day, in their guest room across the street from M.D. Anderson, they waited nervously for their doctor's call. Odegard finally decided he had to call his secret number and get it over with.

"We were all there," says Stephanie. "My mom said, 'Oh John, don't call.' She was very nervous. I remember saying, "Dad don't call, don't call…""

But Odegard made his call.

Stephanie remembers watching her father listening on his phone and showing no emotion. A heavy accent from the recorded voice on the other end of the line made it difficult to understand. Odegard was listening hard.

Then he said, "Uh oh."

"You could tell what it was," she says. "He's like, 'Oh shit, shit…' The voice was talking about spots. It was just an awful way to hear it. But it tells a lot about how he always took control of things."

The spots turned out to be another brain tumor.

"It was a blow of a great magnitude," says Diane. "Plus, Grand Forks was flooding and our house had damage. We immediately started walking around the hospital grounds and down by the river. Then I was on the phone to someone from our prayer group. We were just doing what we could do to get through that news."

They stayed in Houston long enough to have the new brain tumor removed. Meanwhile, because of the flood, the entire city of Grand Forks was without power or water. It had effectively been shut down. UND's President Baker decided to close the university for the rest of the semester. That meant a loss of several million dollars to CAS, because students wouldn't be able to fly, and therefore wouldn't have to pay flying fees.

As the Odegards made their way to their Minnesota lake cabin to recoup, none of that seemed very important.

Oh, I have slipped the surly bonds of earth
And danced the skies on laughter-silvered wings;
Sunward I've climbed, and joined the tumbling mirth
Of sun-split clouds — and done a hundred things.

James Gillespie Magee, Jr.

High Flight

Chapter 27
Those Surly Bonds of Earth

The truth is, by the time John Odegard fell ill, his once humble organization had grown to the point where it was just too big and far-reaching to be run out of the founder's hip pocket any more.

"When the pioneers came out west," says Al Palmer, "they made something out of nothing. Then the settlers came out, and they built on that. When John was here, we were in that pioneer mode. Now we're in the settler mode, and we're still building."

Most successful organizations that start out as the vision of a single leader face the kind of critical juncture that John Odegard's aerospace school faced in the mid-nineties. At some point, future growth and stability demand a change in *modus operandi,* a logical shift from the single-minded individualism of the pioneer to a distributed power model that emphasizes broader organizational responsibility for decision making.

"John was the visionary," says Bruce Smith, now the lead settler at the Odegard School. "He was an authoritarian leader, the hub; things went out on spokes from the hub and came back to the hub. John kind of ran everything from his own realm. As the school got bigger and bigger, grew more diverse, developed more contracts, it became too much for one person. You had to have a distributed organization. But John wasn't quite at that point yet. For an entrepreneur who builds this kind of environment, it's very difficult to change."

Especially while fighting The Dragon.

"John had to be under incredible stress when this happened," says Smith. "Things started to come apart when he was sick. You can see it in the numbers. Everything went downhill at the same time."

Many, including Smith, believe that the turning point in the fortunes of the school during the nineties—and in the stress levels in Odegard's life—actually occurred three years *before* his cancer diagnosis. It was in 1992 that Tom Clifford retired as president of the university, although he stayed on as CEO of the Aerospace Foundation. Following Clifford as president would be a tough, nearly impossible act to follow. The man who succeeded him as president was Kendall Baker, an outsider who had no previous ties to the school or to Odegard. The impact on Odegard of the difference in styles between the two presidents was immediately apparent.

"Kendall Baker wouldn't bail John out, and he was reluctant to let Earl Strinden bail him out," says Smith. "The only thing left was Tom and the Aerospace Foundation. A few years later, when the foundation also started having problems, the whole thing started to unravel."

In a way, Odegard's past had caught up with him at precisely the wrong moment. He'd often spent money from a new contract to pay the remaining costs of an old contract. He'd been lucky, because he kept landing new contracts. But when Clifford left in 1992, and Kendall Baker arrived—on the heels of that Gulf War recession—enrollment in the undergraduate CAS programs remained down. At first, the deficit that CAS faced was also a deficit the university as a whole faced, and even the nation. Gradually, though, the country and UND began coming back. But hiring in the aviation industry remained slow and undergraduate enrollments at CAS continued to drag.

Those slow financial times turned into an official CAS deficit in 1994. It first played out as an annoying buzz temporarily forcing more balls into Odegard's crowded juggling pattern.

"It was just another challenge," says Terri Clark. "We'd sit in meetings a lot of times where he'd say, 'We're surrounded by the enemy, so we can attack in any direction.' Or he'd say, 'One elephant at a time' It was like a call to arms. It was us against the world. He thought we could do it. Did his work make him sick? I don't know."

What is known is that the deficit at CAS never went away while Odegard was alive. Even the Aerospace Foundation felt the effects. Efforts to expand the *ab initio* training to places like India and Kuwait didn't work out. The China Airlines deal had begun to look shaky. By 1998, the year Odegard died, the Aerospace Foundation had racked up its own deficit of about $2.5 million.

Two years earlier, in the fall of 1996, when Odegard was telling the world he'd been cured of his cancer, he flew to England for the Farnborough Air Show. While there he held a press conference, announcing a deal between UND and McDonnell-Douglas. The school was to provide factory training for the helicopter pilots and mechanics of McDonnell-Douglas at the company's Arizona helicopter facility. Many saw it as a much-needed deal. But three months later, McDonnell-Douglas was acquired by Boeing. Within another few months, the deal had fallen apart.

But not everyone was disappointed.

"McDonnell-Douglas had been doing its own factory training, and they weren't making money," says Terri Clark. "They tried to make us think we *would* make money. But to perform this contract, we were going to have to buy the McDonnell-Douglas helicopters at retail prices. I didn't like that. Not all of the ideas we pursued came to be. They were not all good."

Even though Odegard was back at work fulltime that fall of 1996, his illness was an understandable distraction. He had Warren Jensen spending much of his time researching every possible cancer protocol that existed, while he reassured his staff that his illness was "just one more elephant."

"He was always looking at potential cancer treatments," says Jensen. "He wasn't panicky, but he was always tuned into something new."

When he wasn't reading about cancer, Odegard was on the phone trying to put more deals together to bail CAS out of its jam. "I think John managed his stress by planning and looking ahead," says Jensen. "He suffered a lot. But John knew exactly where he was at. He could either dwell on it or work on something that interested him."

Even so, it was clear he couldn't focus on work the way he used to.

"He was at a disadvantage," says Jensen, "because he wasn't here as much. He hired some people I know he wouldn't have hired had it been the normal John. That normal John would have spent more time understanding how a person might fit into the organization. I think he was so distracted he wasn't as thorough as usual."

And because he was away from the office so often, Odegard more or less surrendered some of his authority to George Seielstad, his associate dean.

"I would keep him informed, but he still made a lot of decisions," says Seielstad. "I would be doing more and more and be telling him what I was doing. And then I would only tell him the big things. And eventually, unofficially, I was sort of acting as dean. He had bigger things to worry about."

After the Odegards returned to their home in Grand Forks following a surprisingly pleasant summer of recuperation at their Minnesota cabin, John was dealt another blow that would keep him out of the office even more. He'd embarked on an experimental chemotherapy program with a doctor in San Antonio. In September, John and Diane flew down to San Antonio for a checkup. They were stunned to learn that new tests had shown a spot on John's lungs.

"It was a terrible, terrible blow to John," says Diane, especially after such an optimistic summer. "John knew what it meant. I think we both broke down for just a minute after the news. We felt devastated in the car driving back to Houston. The first hour was bad. There were a few tears in the car, but just momentary. We talked about plans for seeing the doctor in Houston the next day and dealing with it. That night we had dinner in Houston and a glass of wine. I remember that seemed like a nice dinner. We dealt with it. The docs there always had some way of saying, 'Well here's what we're going to do.' And once someone says here's what you're going to do, you're onto another plan and things don't seem so bad. With no plan, you're lost. But we always had a plan."

Odegard went back on his old chemotherapy program and struggled to get into the office. But many knew he had changed.

"He'd work for awhile and be on chemo," says Dana Siewert, "and you wouldn't see him for two weeks. Then you'd see him and each time it was like he was aging weekly, like we age yearly. He was going from 57 to 58 in a month. A few months later, he looked like he was 60. Months later he'd become frail."

But always upbeat.

"What he portrayed to everyone around him was unbelievable optimism," says Corky Everson. "The odds of anyone living that long are so small. Beating those odds the way he did is an amazing story, and it's very much because of his tenacity. He had all those surgeries and chemo with everything in it but the kitchen sink. But John never was pessimistic about anything."

Chuck Kluenker, the current Chair of the Aerospace Foundation and father of two graduates of the program, believes Odegard saw his illness as an opportunity to lead by example.

"John saw he had one more mission," says Kluenker. "The way he handled his illness reinforced to everyone what he and the school is all about."

Eventually, the cancer progressed to the point where Odegard was unable any longer to go out on the road promoting the program.

"What frustrated me the most," says Bob Reis, who was running the foundation in those days, "is that we had some people out there come to us and say, 'Hey, our corporate headquarters is trying to take advantage of you. Walk away from the project.' That was a little tough. I needed to tell John some of these things we shouldn't be involved in. It was tough for him to listen."

In the midst of all of that, the golden deal with China Airlines began going sour. It happened gradually. Under the terms of a five-year contract the airline signed in 1992 with UND, it was to send 36 *ab initio* or basic Spectrum students per year to Grand Forks in two classes of 18 each. As of March 1997 they had only sent five classes.

Through the life of the program, China Airlines continually had asked for upgrades or renegotiated new features as it added new requirements for its pilots. Odegard, of course, had been the driving force behind not only the contract but the willingness to do whatever the airline wanted—or, more likely, to convince them they needed what he wanted. When he became ill, his absence was keenly felt in those negotiations and the normally reliable airline began acting erratically. It would ask for modifications in the Spectrum curriculum and CAS would respond with a proposal. Sometimes they might get no reply at all.

In fact, the pending turnover of Hong Kong from British rule to the Republic of China had caused great anxiety on Taiwan. In March of 1997, says Terri Clark,

As Odegard battled cancer, his attention to the China Airlines business faltered. But the airline had troubles of its own and its connection with North Dakota ended shortly after Odegard's death in 1998. Here, Odegard poses with yet another class of successful graduates of his Spectrum pilot training program for the airline.

the school got a letter from China Airlines saying they wanted to "temporarily suspend" the Jet Spectrum program—for which they had a separate contract projecting training through 2003.

The problem was that Odegard had already budgeted the expected revenue. While some students in the other Spectrum programs did arrive for training the following year, a sense of foreboding hung over the project.

"John was sick," says Al Palmer, "and negotiations were not going well. It eventually went south on us." China Airlines finally negotiated a settlement fee, "It caused some hard feelings," adds Palmer, "because we needed the cash."

This came on the heels of the collapse of the Aerospace Foundation's deal with the Russians to train its Air Traffic Controllers. This deal fell apart due to a different kind of uncertainty: the health of then-president Boris Yeltsin. The foundation's contracts were written so that it purchased services and goods from the university and from CAS. The loss of revenue from the China and the Russian deals was significant, straining relations between Odegard's office and that of UND's President Baker.

"Because of the deficit," says Terri Clark, "which got up to about $3 million, it got to the point where we couldn't hire anyone, even a student employee, without the president's office signing off."

When he was well, Odegard was able to get along with president Baker, though many believed the president had a difficult time grasping just how CAS worked. It was known that Kendall Baker did not see eye to eye with his predecessor, Tom Clifford. As for George Seielstad, the unofficially acting dean, Terri Clark remembers that "George and President Baker couldn't be in the same room discussing finances without raising their voices."

During the later stages of Odegard's illness, the president's office pressured him to sell the Citation jet that Tom Ryan had donated. "I was in this meeting, and we were talking about trying to sell one of the Citations," says Clark. "John sat at the head of the table and he said, 'We're not going to do it.' And he looked at me and said, 'Figure it out.'"

But the only figuring to be done was the cost of maintaining the Citation in a losing project. While the jet was a favorite of Odegard's, it existed as the principal aircraft used to fly university faculty and officials across the state and region. This was the idea that originated 30 years earlier in Odegard's master's thesis and led, indirectly, to the start of his program. Yet for some years, faculty interest in the service—especially given that they had to reimburse Odegard for flights from their own budget—had fallen. In fact, the Air Service had its own million-dollar deficit.

"It was losing money," says Clark. "We were under such scrutiny with the university about every single thing. I used to try to show them that if you were willing to take into consideration the value of your time saved, then the air service becomes cost effective. If you look at the cost of taking a private jet from here to

Kansas City or buying a commercial airplane ticket, the two are vastly different. But in a university setting, they don't often think about that. That money comes from a different spot. They've got this much money for personnel, this much for travel. And I used to say we were trying to provide a service to the university. So don't blame us, because all you are willing to pay is this much, and it's much too small."

Such arguments could not penetrate already made up minds. The university— the ultimate owner of the jet—insisted that the Citation be sold to pay off its deficit. When it sold, it effectively killed the air service.

Many of the old friends and alums who stopped by to see Odegard after his diagnosis, found it difficult to approach him or to know what to say. One of them was Jerry Murray.

"I was feeling awkward about it," he says, "and I remember talking to Warren Jensen. He said that's a big problem for those with cancer. People stop talking to them, stop hugging them, stop reaching out to them. It's the wrong thing to do. He said be there for him. Touch him, hug him, give him things to work on."

Murray took the advice and became a frequent visitor. He remembers one cold February day in 1996 when he dropped in, and Odegard immediately took him out for a long walk.

"I had dress shoes on," he recalls. "We must have walked two-and-a-half miles out and back on crummy, slippery, icy Grand Forks sidewalks. We came back to the house, and we were halfway up the sidewalk when he turned to me. He said, 'Jerry, we'll have fun again.' And then he went into the house. As if to say, 'We'll go fishing again. And flying again.' I believed him. I thought if anybody can beat it you can."

But by January of 1998, most people who knew Odegard could see that he wasn't going to make it. At the suggestion and influence of Tom Clifford, the state Board of Higher Education decided to rename the Center for Aerospace Sciences. It would become the John D. Odegard School of Aerospace Sciences. In addition, CAS I, the first built of the complex of aerospace buildings, would be named Odegard Hall.

Three months later in April, almost 700 people—including Governor Ed Shafer, Grand Forks mayor Mike Polovitz, Kendall Baker, Tom Clifford, the Rev. Andy Fairfield, Episcopalian Bishop of North Dakota, and dozens of alumni— turned out for the official dedication ceremony. By then, Odegard was seen less and less in the office. But he'd saved up his strength, says Diane, and came out that day and gave a strong speech of thanks and stayed for some of the festivities.

When Jerry Murray, a public relations executive in Minneapolis, learned that the school would be renamed for his mentor and former next door neighbor, he and his partner, Jim Moore, began brainstorming.

Murray knew the ceremony was slated for a Friday afternoon, with a reception on Saturday. He figured the family would be in a quiet private time on Sunday, when all the excitement would have peaked. Likely, John would start to think about cancer again.

"I knew I'd regret it the rest of my life if I didn't get something done on time," he says. Working on a tight deadline, Murray contacted dozens of aerospace industry companies who contributed money and their logos to a unique full page ad that ran in the *Grand Forks Herald* the Sunday following the re-naming ceremony. Murray wrote the copy, which was a tribute to Odegard. The headline that ran above a dramatic aerial photograph of the sprawling aerospace campus said it all: "This is the House That John Built."

"Nobody knew the ad was gonna run except myself and John Junior," says Murray. "Sunday morning came and John's daughter Stephanie went out to get the paper. She and Diane were downstairs at the breakfast table, but John was in bed, not feeling well enough to come down. All of a sudden Stephanie turned the page, and there was the ad. She said 'Mom, look at this.' It did exactly as we wanted. It just lifted them.

It was a special moment for John and Stephanie when she ran upstairs and said to him, 'Look at this, Dad.' It accomplished everything we wanted."

During that spring of 1998, the school planned to take delivery of two new Beechcraft Barons from the company's factory in Wichita. Odegard's son, John, Jr., flew one of them to Grand Forks from Kansas. He'd graduated from his father's program in 1989 and for a time served as a UND flight instructor. After earning a master's degree at St. Thomas in St. Paul, Minnesota,, he took a marketing job in Wichita at Raytheon, the corporate owners of Beechcraft.

His father, he says, never pressured him into flying. But when he started formal flight lessons at 15, his father did give him some instructions. "My father was a very good teacher," he says. "He loved showing you things. He couldn't resist wanting to get into a cockpit and fly. I don't remember him ever saying why he loved it. I just remember him being absorbed in it. It's what made him a good teacher. It wasn't just something he flew, but how it flew, what made it fly and what caused it to turn. The mechanics and aerodynamics and physics of it all. He was intrigued by that. He said, 'I'm a pilot.' That's how he classified himself. That was the number one passion in his life. He always flew. He would not have said he was a professor or even a businessman. He might say he worked at the University of North Dakota, but he never presented himself as the dean. He had great humility."

John Jr., made his first solo flight on his 16th birthday. Immediately afterwards, he took his father up for a flight, his first ever passenger. Then, on that spring day in 1998, Odegard senior was at the Grand Forks airport to meet the Beechcraft Baron his son was piloting. Odegard had assigned himself one last duty of performing a check ride in the new aircraft. Local television crews and newspaper reporters were on hand to record the event. Odegard chatted with them about all the positive happenings at

Following the ceremony renaming the school in his honor in April, 1998, Odegard relaxed
at home with his family. From left Diane, John Jr., Odegard, his daughter-in-law Paula
Odegard and daughter Stephanie. He succumbed to cancer five months later.

the school that now bore his name. He would say only that he hoped the name change would help the school.

When John Jr. set the Baron down in Grand Forks, his father climbed aboard, taking the unfamiliar right seat. Because of his illness, he'd lost his medical clearance to pilot an aircraft. With that, father and son took off. It would be Odegard's last flight in a UND aircraft.

The last time Odegard flew at all was for a final checkup in Houston. In May, when he and Diane were back in town, Jerry Murray flew into Grand Forks for a visit. Odegard's secretary, Jennifer Foss, picked him up at the airport.

"I asked her, 'What are John and Diane doing today?'" he remembers. "And she told me, 'They're at the funeral home.' That's when I knew for sure that they knew they weren't going to recover from it."

Indeed, says Diane, when things started to go down, John started planning his own funeral. He picked out his own casket, selected a plot of ground in Memorial Park cemetery and authorized work to begin on a small mausoleum. "He was a pilot," says Diane. "He wanted to be above ground.

"All during that summer we would drive out there and see how it was coming. He'd drive the jeep right up to the point where it would be. It was important for him to plan and to be in control. John said, 'Well, you might as well have a nice spot.'"

In June, he stopped going into the office, making one last visit in July when his friend Ken Svedjan and others planted a tree in his honor outside of Odegard Hall. Then, one July afternoon while he and Diane were at home, sitting in their backyard, he started telephoning friends and alums. He was lining up his pallbearers. Jim Bunke remembers that phone call very well.

"John said, 'I've got an important request.' I said, 'Fire away.' He said, 'Would you be a pallbearer?' And you're just kind of numbed by the question. And it's like 'Okay, Bunke, now what do you say?' So I said, 'Of course I will, but I hope you're a fat ass old man by the time I do it.' He chuckled and said, 'Thanks, Bunk.' And that was the end of the conversation. My wife said, 'What did he want? I said, 'He wants me to be a pallbearer for him. Wow.'"

Later that summer, Odegard worried about where the post-funeral reception would be held. The basement of St. Paul's Episcopal church was still being repaired after the flood of the year before.

"I know," he told Diane, "we'll have it at the school!"

On September 12th, Diane threw John a birthday party. He was 57. By now, Odegard was confined to his bed upstairs, under home hospice care. One by one, the guests at the party made their way upstairs to sit with him and say their hellos and goodbyes. Diane checked in at one point, and found Tom Clifford sitting on the end of the bed rubbing John's sore feet.

"And Tom was saying to John, 'Now, I've found this works really well...' That

When the university renamed CAS the John D. Odegard School for Aerospace Sciences in 1998, many of Odegard's graduates returned for what would be a final visit. Here, Jean Haley Harper, a United Airlines captain, visits with a frail Odegard after the ceremony.

was the last time Tom was with him. I think they said their goodbyes."

The next day, Diane drove John to their Minnesota lake cabin for one last visit. When they returned home, they decided John should be moved downstairs into a hospital bed. Jim and Cari Odegard were there to help

"But John decided he was going to walk down the stairs himself and get into the hospital bed, which he did," says Diane. "And of course he never got out of the hospital bed."

A television was perched nearby as was his organizer, a laptop and his wallet containing all of his credit cards. "On TV one day, he saw an ad for this broom that would expand and you could clean the ceilings," says Diane. "He ordered that for me over the phone. He was a great orderer of things. And that was my last present from him."

John and Diane's high school friend, Del Rae Meier, visited him in those final days. A self-described "religious person" Meier says she was always amazed that John could recite from memory the prayer he had chosen as a boy for his confirmation.

"By now his friends had seen him in a whole new light," she says. "To see him being vulnerable and reaching out and being open. That was not his style. But he was okay with it. John just took things on faith. He was an action person and flying was his spiritual yearning. His getting closer to God."

Two weeks after John's birthday party, Del Rae stopped in to rub his feet. He was unable to speak. The next morning, the 27th, she got a phone call from Diane. John had died early that morning.

He'd asked that his confirmation verse—"For without faith it is impossible to please him…"—be inscribed on the face of his tomb.

"When I first came here," says George Seielstad, "people didn't like the Odegard School. Envy or something. He did everything a different way. He thought, 'We've got to get things done.' What happened when he got sick—and it was evident he was in a terminal phase—all of a sudden people said, 'God, look at what that guy's done.' The rest of the campus began to see 'Well there really are academics over there. It's not really a trade school.' Anyone can learn to fly, but John thought this should be a profession, like accounting or law. You should have some liberal arts and a business background. You need to understand other cultures, so you should learn a foreign language. You should also become very skilled at the aviation part and the profession. But it should be in this broader context. They missed what John was trying to do. The poor guy, when he died, I think it finally sunk in."

PART FIVE
HIGH FLIGHT

No bird soars too high,

if he soars with his own wings.

...reach for the heavens,
and hope for the future,
And all that we can be,
not what we are.
John Denver
The Eagle and the Hawk

Chapter 28
I Am Not John Odegard

Aand so began the search for John Odegard's successor. Recognizing the likelihood that the John Odegard mold had been broken soon after his birth, the university search committee decided, wisely, not to waste time searching for a clone.

"We knew going in that we weren't going to find another John," says Bob Buley, a committee member. "We were looking for some element that would at least fill the gap. We looked hard, high and low."

They looked for someone who, if not exactly John Odegard, would still feel a positive resonance from the hundreds of classic case studies in the Odegard pantheon.

For example, Al Palmer recalls a meeting a few years after John Odegard's death with a group of leaders connected to The Ohio State University's aviation program. They had come to Grand Forks to find out what it would take for them to duplicate the success of UND Aerospace. Among them were members of The Ohio State University board, along with a senior vice president of NetJets, named Richard Smith. Based in Columbus, NetJets essentially leases corporate-style jets to companies by the hour, saving their customers the millions it would cost to buy a jet outright. NetJets, in fact, pioneered what is now a booming business called fractional ownership.

After leading them on an extended tour of the facilities, including a visit to Odegard's grave, Palmer remembers Richard Smith saying that NetJets was going to buy planes and build buildings at Ohio State and was going to make it "just like UND."

"I said, 'Richard, it's not going to happen.' He said, 'Yeah it will.' I said, 'No it's not. And the reason it's not going to happen is that you can't buy people. What makes a hearty organization—our best resource—is people. You can't buy them."

"Sometime later I was back, talking to Richard," Palmer recalls, "and he admitted I was right.' Since then we've had other colleges come up here and benchmark. We'll share our best practices with them. But if you don't have that drive, if you don't have John Odegard—if it wasn't for John this would be the little flight school on the prairie."

Then there's the conversation Warren Jensen had a while back with an economics professor across campus.

"I said to him, 'Gee we're working on a master's program,'" recalls Jensen. "'Would you like to come and teach airline economics? We can help you get all the airline information.' He said, 'No I'm already assigned nine hours to teach, I don't teach more than that.'"

Jensen said no more, seeing it as a losing cause. In addition to his duties supervising the altitude chamber, he taught 15 hours a week—12 on them in undergraduate and three in graduate classes.

"There's an excitement out here that's different," Jensen says. "We have this need, this panic to succeed. John always felt we've gotta do this to be the best, in order to make it. We don't ever say it's not in my job description, or it's more than I can take, or my plate is already full. Compared to across campus, it's a totally different mindset. They look at us like 'You guys just don't say die over there.' We make them tired."

The search committee, chaired by Dennis Elbert, the dean of the business school, numbered several aerospace faculty and staff as well as representatives from the aviation industry. Among them were John Odegard Jr., then a sales executive with Raytheon's fractional ownership program. He is currently a vice president with NetJets. Other committee members included Bob Buley at Northwest and Jim Bunke, then at Beechcraft in Wichita.

"A fair number of us had emotional ties to John," says Bunke. "There was an unusual number basically involved in hiring their future boss. It was highly emotional. We had to start with the premise that there is no other John Odegard. So you say, okay, who can at least keep the program viable? Someone who will love the place and recognize the vision and build on it. We had a hell of a time."

By April of 1999, the committee had whittled its list of candidates down to eight. Among them were four of Odegard's own hires over the years. They included George Seielstad, the associate dean; Leon Osborne, director of the regional weather information center; Charles Wood, the chair of space studies; and Richard Nelson, the managing director of aviation who had been named interim dean shortly after Odegard's death.

Nelson, the former president of the Milwaukee-based regional carrier Skyway Airlines, had been hired by Odegard in 1996. In addition to the title of interim dean, he was also named interim chief operating officer of the UND Aerospace

Foundation.

The other four candidates included a professor of psychology at Central Florida University, the commander-in-chief of the 119th fighter wing at Fargo, the manager of space environment systems at the Teledesic Project, in Tempe, Arizona and the director of training at Delta Airlines in Atlanta.

That list was further culled to three finalists: but one name among them stood out. Bruce Smith, the Delta executive had caught everyone's attention from the moment his application letter arrived. He was a 1970 UND graduate with a double major in mathematics and education. While at UND, he lettered in track and football and was named to what was then called the "Little All-America" team. Smith had also taken a ground school course from John Odegard himself.

A former Air Force jet pilot and member of the faculty at the Air Force Academy, Smith earned a Ph.D. in Instructional Design at Florida State University. Later, in a variety of industry positions, he helped design a simulator upgrade for the space shuttle, and simulators for the B-2 bomber and the F-117.

"Bruce rose to the top right away," says Bunke. "The two others were second and third but a long ways from number one."

In his letter of application, Smith wrote, "This is the position that I have always viewed as my career ideal. At times I have even dreamed about serving in this capacity, while at the same time I was certain, and later sincerely hoping, that it would never be open during the span of my career." Because, Smith explained, if it did come open, it would mean something had happened to John Odegard.

"I'll never forget reading Bruce's letter," adds Bunke. "All of sudden we're thinking, 'This guy's got the emotional tie. He has incredible loyalty. If they hadn't offered him a football scholarship at UND, he may never have gone to college. We were clicking along."

But then, citing personal reasons, Bruce Smith suddenly withdrew his application. At that point, says Bunke, the committee chair wanted to wrap the search up and forward to the president the names of the two remaining candidates.

"I remember saying, 'Guys this is not the right thing to do,'" says Bunke. "The other two candidates were so far below what we wanted."

At an impasse, the search for a successor to John Odegard was suspended indefinitely.

In the meantime, the Odegard School forged on. Funded by a $1 million donation from Tom Ryan, the walkway across 42nd Street—Odegard's so-called gerbil tube—connecting Clifford Hall with CAS III, was completed and dedicated. While Odegard had frequently visited the construction site, he never did get to see it entirely completed. Later, in recognition for his donation and years of support, CAS III was named Ryan Hall.

At about the same time in the fall of 1998, the long and painful deficit that had caused so much stress at the school, gave way to Odegard's hard-working staff and a national economy that finally showed some strength. Undergraduate registration at the Odegard School was up by 15 per cent. In fact, with a fleet then numbering 100 aircraft, the school was looking to hire 30-40 flight instructors.

"In my mind, the enthusiasm from key employees was still there," says Smith's associate dean, Paul Lindseth. "We still carried with us John's vision. We still had the motivation to move forward."

Yet in some areas at the school, there seemed a lack of energy and growing uncertainty. In addition to the disruption of Odegard's death, a new president, Charles Kupchella, had taken over on July 1, 1999. No one was sure how he viewed aerospace or what the future would hold. In addition, there were several acting vice presidents among the university's administrators, causing Odegardian worry over a perceived lack of aerospace empathy.

"Our problem was hesitancy," says Warren Jensen. "John didn't hesitate. He went into something and did it. Suddenly it was, 'What do we do?' We were so not-wanting to make a mistake that we became hesitant. We were adrift for about two years because of that uncertainty. We still did what we did. But before, we were all working together. John could keep the different factions integrated. When that stopped, we all continued to work well, but it was kind of like the ropes between us were broken. We continued to work and do our own thing, but we didn't integrate as well. John had always taken care of that for us."

While others suggest this malaise was not an organization-wide problem, outside friends and alums of the program, such as Bob Buley and Jim Bunke, felt the program wasn't nearly as vital as it had been. They remember well that frustrating meeting of the search committee where the successor search was put on hold.

"We'd spent a frustrating day talking about it," says Buley. "Having come to the conclusion that the only candidate we wanted had decided he wasn't going to take the job—Bruce Smith—and we were going to have to go back to the drawing board and start all over again. Bunke and I were walking out of the meeting, and I said, 'You know what Jim, let's go to Odie's grave.' So we drove over there. I don't have much of a feeling for graves and all that stuff. We just went over there. When we got there, we said, 'All right what would Odie do?' We decided Odie would go down and see if he could convince Bruce to change his mind. So that's what Bunke and I did."

But not before Bunke raised a delicate legal point, "Under North Dakota law, all of our dean search meetings had to be public," he says. "I said, 'Bob, there's a little risk in that.' After a pause Buley says, 'So what do you think Odie would want us to do?' And I said, 'What time is the flight, Bob?'"

Neither of them had met Bruce Smith before, but he agreed to meet with them.

"We were really impressed with Bruce," says Bunke. "We had done elaborate interviews with people who knew him. We learned a lot: he was articulate, he wrote very well."

Smith explained to them his reasons for withdrawing. Both his sons, following in the old man's footsteps, played college football. Allan, the oldest, had played for Angelo State University in San Angelo Texas, graduating in 1997. His youngest son, Jay, played at the University of Oklahoma where he was then entering his senior year.

Smith had never missed any of his sons' games. Knowing the culture at UND, and the 100 per cent commitment the dean's job would demand, he thought it wouldn't go over well if he took the job and then disappeared for 11 consecutive 4-day weekends to watch his son play for OU.

He told Bunke and Buley that when he applied for the job, he thought maybe things could be worked out, perhaps allowing him to start on January first, after football season. But the search committee had been quite firm about filling the job by the first of July.

Bunke and Buley thought there might be room for negotiation there, but there was another issue. Smith's wife, Ann, was not completely convinced about the move.

Bunke remembers saying things like, "Could we have lunch with Ann?" Or "How about we have a drink with Ann? We could swing by the house and get her?" Smith was reluctant to subject his wife to the man whom John Odegard had considered his protégé as a master salesman.

"It became obvious we weren't going to get to meet Ann while we were there," says Bunke. "So, it was hotter than hell in Atlanta. And I said, 'Bruce, we know of a place where it wouldn't be nearly this hot.' At that point Buley nudges me and says, 'Lighten up a bit, Bunk.'"

They flew back to Minneapolis more impressed than ever with Smith but feeling they had not moved him an inch off his decision. In fact, their impromptu visit had made a strong impression on Smith, enough that he and Ann talked about going out to North Dakota in the fall for a Homecoming visit.

In the meantime, Bunke and Buley bumped into Earl Strinden in the Minneapolis airport. As Bunke recalls it, "Strinden says, 'Jim what brings you to the airport today?' I'm kind of looking at my shoes and I finally confess. I said we'd just come back from an illegal recruiting trip to get Bruce Smith to be the dean. Earl looks at the two people with him, and he says, 'Gentlemen, *this* is dedication!'"

As the search continued, it became evident that the July 1 start date for the new dean was going to slip closer to January 1, 2000. In mid-October Tom Kenville, Jr., the marketing director of the Aerospace Foundation, flew to Atlanta to attend NBAA and while there arranged a visit with Bruce and Ann. During that visit,

Bruce Smith's first message to the faculty and staff at The Odegard School when he took over as dean was "I am not John Odegard." Yet his strong academic credentials and solid management style have kept the school advancing in Odegardian fashion, with the spirit of the founder very much in evidence.

What started out as an unsightly Quonset hut barn in 1967 with three airplanes had become a modern airport facility by 2007. The UND operation at Grand Forks Airport now numbers several hangars, three ramps, an office building, and multiple jets and helicopters.

Kenville indicated that the search committee was still undecided about its final candidates and the date to fill the position could easily be delayed to January. For Smith, that removed a key barrier. The only things that remained were an informal visit to the campus to help convince Ann that this was the right decision and to meet the selection committee for the formal interview.

As it turned out, UND's Homecoming weekend in the fall of 1999 was a bye week for Oklahoma's football team. The Smiths were able to fly up to Grand Forks and tour the campus with Tom Kenville. The Smiths were hosted by Tom Clifford and his wife Gayle, and even had the opportunity to meet North Dakota's governor, John Hoeven.

"That's when I began to realize," says Smith, "that the planets were aligned here. It was also important for Ann. She didn't realize the extent of the aerospace school."

Bruce and Ann both grew up in St. Louis Park, a suburb of Minneapolis. They'd attended the same elementary and high schools. Bruce had first asked Ann for a date when he was in fifth grade; her mother said no. But they dated through high school and into college. Bruce went to UND on a football scholarship—the only way he could afford an education—and Ann went to the University of Minnesota.

In his sophomore year at UND, Smith was injured in a pre-season football game. It's a moment he remembers now as one of those teetering fragile instants when his entire future hung on what he did next. He asked the coach if he could skip that coming sophomore season on the team and start over with his eligibility intact the following year. In doing so, he became one of the first of a growing college trend of "red shirting" student-athletes.

But at the moment, because of the Vietnam War, Smith was pretty sure he would be drafted if he didn't complete his draft-deferred college education in four years. A moment from his boyhood bubbled to the surface—a day as a fourth grader, long before he thought of dating girls—when he heard his first sonic boom. Looking toward the sky he'd said to himself, "There's someone up there going awfully fast. I want to do that."

So with red shirt in hand, he went to the university's Air Force ROTC office and found he could be guaranteed a spot as a pilot trainee if he joined the program. Which he did.

In the summer of 1969, Bruce and Ann married following her graduation from Minnesota. She moved to Grand Forks that fall and taught school at the Air Force base.

Bruce, holding down a double major in mathematics and education, was the starting center for the university football team. It was also the year he met John Odegard, whose fledgling aviation department had won a contract from the Air Force to give ROTC cadets a flight indoctrination course.

Smith took the ground school course from Odegard that fall. He did his actual flight training over the winter, graduating with his private pilot's license.

That year in Grand Forks was Ann's only experience with John Odegard and his aviation school. Thirty years later, when Bruce told her he was thinking of applying for the dean's position, she was puzzled. She remembered the two or three planes Odegard had, the drafty old Quonset hut at the airport and the crowded offices in Gamble Hall.

"So that Homecoming was an important visit for Ann," says Smith. "She didn't realize the extent of the school here. She didn't know about the four major buildings and the changes at the airport."

The Friday after Homecoming, Smith was back in Grand Forks, officially interviewing with the search committee. That night in his motel room, he got a call from Charles Kupchella, UND's new president of six months. He was offering Smith the job.

Without question, any outsider who tried to replace John Odegard was going to be viewed skeptically and with suspicion, not only by the aerospace school but the suspicious skeptics across campus. Following Odegard was a nearly impossible challenge, as difficult as following Tom Clifford had been for Kendall Baker—who never developed the respect of his predecessor. It was a tough act for anyone except, perhaps, Bruce Smith.

He did have the perfect credentials. He'd been a jet pilot with the Air Force, where, in a T-38 jet, he'd fulfilled his childhood dream of reaching supersonic speed and breaking the sound barrier. He'd been one of Odegard's students. He'd been a football All-American in a town that loved football almost as much as hockey, he was a director at a major airline, he knew hardware, he knew software and yet he was also an academic with scores of published articles to his credit.

"I think there was an expectation that I was the right guy," says Smith. "If I'd come in with anything else, there would always be some question of whether I was the right guy."

Even so, he felt nervous because he was a UND grad and had so much at stake personally in succeeding.

"I'd had four major leadership positions in my career until then," he says. "Two were very successful, two were less so. You learn from failures. Introspectively you look at this job, and you see that this isn't something you want to step into and fail. You can't mess this one up. There's too much at stake."

Actually, Smith unveiled one more credential the very first time he met with his new colleagues at the school. He told every one of them what they wanted to hear: I am not John Odegard.

"How do you step in behind a legend?" asks Smith. "It would have turned a lot of people off very quickly if I tried to come in and make this my place. So I said we're going to continue to revere John. He's always in our hearts and on our minds."

Even before he took the job, Smith had sought the advice of Jeff Lickson, a close friend from the same Ph.D. program at Florida State. He'd become a top management consultant, specializing in organizational behavior and development. Smith wanted to work out a rational plan for handling the key issues he'd be facing. They ranged from dealing with the various personalities, the longstanding cross-campus acrimony, the need for building the academic prestige of the school, and knocking down some of the silos of special interest within the school—especially between the aviation and non-aviation components.

One thing Smith wanted to do right away was get working on a 10-year strategic plan. But he wanted to avoid what he saw happen with the drafting of a strategic plan at Delta. In that case, he says, key people were brought together at a convention center and worked 12 to 14 hours a day over the course of four days putting together a plan. "People got tired and upset," Smith recalls, "and dominant personalities started to prevail. They came out with their plan, and we looked at it and said it was nuts. But they implemented it anyway."

So when he took over, Smith brought Lickson in for several days a month to coordinate with the staff and faculty the steps necessary to develop a 10-year-plan.

"He told me that I tended to be a problem solver," says Smith, "but that the plan had to be a team project."

Smith knew all about team projects from his days as an All-American football player. The legend on his plaque in the UND Athletic Hall of Fame reads, "Being an offensive lineman you expect to get hit on every play, get beat up, dirtied and bloodied. You don't expect a lot of recognition yet take great pride in your contribution to someone else's success. You truly understand the meaning of the word teamwork."

Lickson conducted dozens of surveys on campus and in Grand Forks. He interviewed college staff, faculty, the president, the chamber of commerce, the city council and everyday citizens, asking their expectations and perceptions of the aerospace school.

"It allowed me to fully devote time to running the school and establishing things in terms of goals," says Smith. "I wouldn't have time to do both."

In the meantime, Smith had taken Lickson's advice. He worked through Odegard's old "dean's staff" organization—renaming it the "leadership team"—and tasked them with devising the 10-year plan. The vision of the organization now lay in the hands of those 22 team members—including Smith. At first, Lickson would work with team members individually, finding where they personally wanted their department to go. Then they worked as a group to find out where the group as a whole wanted to go. He returned to UND each month through March 2001 to keep ideas for the plan percolating.

Nine months after they'd begun, in October 2000, a 10-year strategic plan, agreed to by all, was in place. If nine months seems a long time, says Smith, at least no one said the resulting plan was nuts.

"We wanted to give people time to reflect and contemplate their future," he says. "You can't rush that process."

Since then, the leadership team has remained in place. In fact, every goal on the ten-year plan was accomplished in five years.

"It's been an incredibly helpful group," says Smith. "We have some heavy duty discussions about where we should go with things. It evolved into a problem-solving group. I don't go in and say, 'Here's the information.' It's more like, 'What's everyone working on and what help do you need?' Then the group as a whole helps. We were able to break down those silos because everyone knows what everyone else is working on."

The difference in management styles was evident to all.

"John was all about 'Do it now,'" says Warren Jensen. "Bruce is a planner. John was the entrepreneur, Bruce is a manager. He did a phenomenal job when he came in. He said, 'I want to know this organization.' We spent months forming leadership teams, having discussions, filling out questionnaires. He knew he wasn't going to mimic what John had done, but he fostered it with a concentrated, clear effort to understand us and to lead us. He understands where he needs to put efforts and where he does not. He steps in when he needs to but not in a heavy handed way. John was hands on and Bruce is hands off."

Smith is so hands off that he did something Odegard couldn't bring himself to do: he shared power.

"Because John had been ill and had spent a lot of time trying to just get well," says Smith, "there was a fend-for-yourself approach here. John started the school when he was very young. He had brought a lot of young people along with him. They grew up together, and they had developed almost a parent-son-daughter relationship with him. When I came in, they were ready to take over the leadership roles. Psychologically, there was a major shift.

"What happened is they responded to the empowerment. They were ready to take over. They said I don't need the direction, the oversight, the micro management. I know where to go and what to do. And they did. Other places may have been threatened by this, but they weren't. They no longer needed nor wanted authoritarian leadership. It became a distributed organization all on its own."

Each time we make a choice,
we pay with courage
to behold the resistless day,
and count it fair.
Amelia Earhart

Chapter 29
The Once and Future Odegard

O n a Monday morning in October of 2000, nine months after Bruce Smith had been named the dean of the Odegard School, he got the kind of bad news an aviator dreads. One of his flight instructors had died in the crash of a university plane at the airport in Rapid City, South Dakota. Worse, it appeared that the instructor, flying alone, had intentionally wrecked the plane. The ruling was suicide.

No matter how distributed the power gets in an organization, when such moments arise everything falls back on the man in charge. The death of Bob Thompson was keenly felt by students and instructors for he had been extremely popular. Smith felt an extra measure of pain. Thompson had been a bright undergraduate at the Odegard School who had an alcohol-related offense on his driving record. The faculty of the aviation department had worked hard to help Thompson overcome the stigma of the blemished record with regard to the no-tolerance policies of the commercial airlines. He'd gone through a rehabilitation program, earned his pilot's license and had graduated with a 3.7 grade point average.

Smith recalls the young man as "passionate, hard working, an exemplary instructor." He was aware that Thompson had wanted badly to become an airline pilot, and knew that the aviation department had worked with him on practice interviews. Only a few days before the fatal crash, Thompson had interviewed with a regional airline, but had done poorly.

"He came back that weekend," says Smith, "and was charged with his second DUI. That Sunday he sat down and wrote goodbye letters to his family. On Monday, he checked out an airplane and flew to Rapid City. He did a couple of touch-and-goes. He got on his radio and said, 'Call my parents and tell them I love them.' He then dove straight down into the center of the runway."

In Grand Forks, emotion ran high. Other instructors and student pilots immediately wanted to check out planes and fly to Thompson's home near Oshkosh, Wisconsin for the funeral.

Smith said no way. Flying with that kind of emotion clouding ones' thoughts was an invitation to disaster. Instead, he persuaded the university's athletic director to loan him the use of the team bus. He loaded it up with all of Thompson's friends and drove to Wisconsin. Benefactor Jim Ray, who had years earlier funded a Young Eagles retreat lodge at the headquarters of the Experimental Aircraft Association in Oshkosh, arranged to house the busload at the lodge overnight. Unlike at a hotel, there was plenty of room at the lodge for everyone to hang out and reflect, says Smith. The next day after the funeral they all rode the bus back to Grand Forks.

But that wasn't the end of it for Smith.

"We looked at our program," says Smith, "and asked what do we need to do to prevent this? We decided we had way too much emphasis on the left-seat career in a big airplane. That kid was so focused on being an airline pilot that when his dream didn't come true, there was no place to turn. But of course, there were thousands of places.

"So we started to re-emphasize other non-piloting careers as much as commercial aviation. We brought in lots of people in aviation who were not flying big airplanes, people like Jim Bunke and Jerry Murray."

Smith also put all of his instructors through suicide prevention training. Two years later, he got a phone call from a young woman who'd been on the bus to Oshkosh and who had taken part in the suicide prevention class. She went to work as a dispatcher for NetJets in Columbus, Ohio and was part of the company's flying club.

"She was flying over Columbus," says Smith. "She heard a radio call that was identical to what Bob Thompson had said: 'Take this phone number and call my parents and tell them I love them.' She calls right back on the mike: 'Who are you, where are you, what are you doing?' It was a 17-year-old kid who'd stolen a plane. He had 11 hours of flying time and he was going to commit suicide. He was 50 miles away. She flew to where he was and talked him out of committing suicide using techniques she'd learned in suicide training. Then the kid says, 'But I don't know how to land.' She says, 'Don't worry, I'm a flight instructor.' And she talked this kid down. The first landing was so far off she had to talk him through a missed approach. She got him down the second time, and he landed safely."

The South Dakota incident demonstrated to the staff and faculty of the Odegard School that the new dean could handle a Monday morning crisis as well as a 10-year plan. With renewed confidence, Smith turned to the list of personal goals he'd set for himself when he first sat down in John Odegard's office.

"None of those objectives," Smith says today, "was particularly measurable, but they were pretty accurate in terms of what needed to be done."

A key idea was to keep the school in the forefront of aerospace education. Two challenges immediately stood out. The financial ups and downs of the school and the Aerospace Foundation were well known, and certainly measurable. But the strength of the academic program at the Odegard School was an amorphous issue, something that had never really been addressed head-on.

"We were accredited and teaching good courses," says Smith, "but it was a perception thing. Parts of the campus still viewed us as a technical school. Our programs needed to be strengthened. To take the school to the next level, you had to have the kind of prestige you get with master's degrees and Ph.D.s.

When Smith arrived in 2000, the department of computer science already offered a masters program, as did the department of atmospheric sciences and the distance-learning space studies program. But no Ph.D. programs were available in any department. The department of aviation, meanwhile, offered several undergraduate degrees, but had no master's program.

One of Smith's first decisions was to appoint Paul Lindseth—the interim associate dean of the previous two years—to be his permanent deputy as the assistant dean for academics. He charged Lindseth right away with developing advanced degree programs throughout the school.

Like Smith, Lindseth is a former Air Force pilot who was mesmerized by flying as a boy in almost identical circumstances. Raised on a farm near Rugby, North Dakota, Lindseth remembers standing in a field and being enchanted by the sound of sonic booms from high flying jets. "I remember looking up and saying to myself, 'That's something neat I'd like to do.'"

The problem was that none of Lindseth's three older brothers wanted to farm. Feeling that someone in the family had to keep up the tradition, he enrolled at North Dakota State University in Fargo. But by the time he'd graduated, his older brothers had seen the light and were already back on the farm near Rugby.

Hedging his bets, Lindseth had enrolled in the NDSU Air Force ROTC program. His dream of flying jets came true, and he spent nearly 10 years in the Air Force as a jet flight instructor. He also earned a master's degree in management.

Lindseth was hired at UND in 1985 as a flight instructor of fixed-wing aircraft, later moving into helicopter training. Promoted to associate professor in 1988, Lindseth taught courses in air transportation and aircraft systems.

During the time that John Odegard fell ill, Lindseth was completing a Ph.D. in higher education and academic administration through the University of Michigan. "John was always interested in what I was doing to get a doctorate," he recalls. "He was extremely ill with cancer, but he'd come up and put his arm around me, and we'd walk down the hall, and he'd want to know everything about how I got this degree. Like he was thinking of doing it himself."

While the army once considered canceling its support of the helicopter training program at UND, it reconsidered when Sen. Byron Dorgan convinced army generals that it was cheaper to offer initial chopper training in Grand Forks so that graduates would immediately join the advanced class at Fort Rucker Alabama, a step ahead of their peers.

Paul Lindseth, PhD, the associate dean of The Odegard School, handles all academic matters. Like his boss, Dean Bruce Smith, Lindseth is a former air force pilot who heard a jet breaking the sound barrier as a boy and promised himself he'd do that one day. And he did.

Lindseth supported Bruce Smith's view that the school's academic credentials needed an upgrade. "The key indicator of a high-quality program is the curriculum," he notes. "Does it respond to the needs of the industry or the academic discipline? The way to do that is to add master's programs and doctoral programs."

Since 2000, a main thrust at the Odegard School has been toward getting advance degree programs approved by the university. Under Lindseth's direction, a master of science program was developed in aviation; plans are underway for a program leading to a doctorate in aviation. In the meantime, the doctoral program in atmospheric sciences has been approved—although not without a good bit of bureaucratic hoop-jumping.

At one of Bruce Smith's very first get acquainted sessions with the computer science department, a case was made for a Ph.D. program in computational science. Smith agreed and with Lindseth, began the groundwork to get the program designed and approved. It turned out, though, that the design wasn't the real problem. The approval process itself became a turf battle typical of university budgetary politics.

The math department in the university's College of Arts and Sciences thought the doctoral program needed more math courses. But they also said they didn't have the funding to offer those courses. There was also a concern raised when North Dakota State University in Fargo suggested that such a doctorate would compete with its own Ph.D. program in computer science. There was even a rumor that should UND go ahead, NDSU might retaliate by offering a competitive Ph.D. in English. It took Smith years to get across the point that computational science and computer science were as similar as football and foosball.

While computer science focuses on theory, computational science is more about the practical application of that theory across the scientific curriculum. It might, for example, help chemists to use the tools of computer science to develop a model for solving a given problem.

Smith admits to frustration with the approval process for new programs which he likens to an illogical circular argument. "One of the things that helps a department, that needs help," says Smith, "is to create a Ph.D. program. It attracts strong graduate students who help improve the department. But they say, 'Well that department might not be strong enough to do a Ph.D.' But that's why we want to start a Ph.D.!"

As those doctorate programs have made their way through the university's labyrinth, Smith has focused on hiring more Ph.D.s or academics with terminal degrees. Several people in the aviation department have also earned their doctorates in part time studies.

In the fall of 2005, the Odegard School inaugurated a Ph.D. program in a brand new graduate level department called Earth Systems Science and Policy. This was one of the spin-offs of George Seielstad's work for NASA that started back in 1994. When he founded the Upper Midwest Aerospace Consortium, he simultaneously

The Odegard school's mainstay student aircraft are built by Piper and include the Warrior, the Arrow and the twin engine Seminole.

The National Aeronautic and Space Administration loaned this DC-8 to UND for use by scientists studying application of data collected on NASA space flights. That program, led by George Seielstad, is unlike any other in the country.

began a research program in earth system science. It was out of the consortium's research—drawing about $14 million in outside funding in its first ten years—that UND Aerospace built its earth systems science program.

In conjunction with UMAC, Seielstad also launched the Northern Great Plains Center for People and the Environment. One of its aims is to increase the science literacy of the public, especially its understanding of the delicate nature of the planet earth. In the meantime, because of the Odegard School's existing array of aerospace programs and its proven aviation professionals, Seielstad found NASA quite receptive to loaning the university a DC-8 jet and providing it with $25 million for sub-orbital research projects.

In recent years, Seielstad's programs have branched out of the Odegard School and now fall under the auspices of the university's vice president of research.

Just as strengthening the academic life at the Odegard School has progressed successfully if not smoothly, the financial wrinkles of days gone by also have been ironed out. In fact, Bruce Smith's solid reputation as a manager resuscitated the school's fiscal credibility almost from the day he arrived.

"I'm much more conservative than John," says Smith. "Financially as well. We've become more fiscally responsible. John got us where we are. Some of it was outrageous. Some of it I wouldn't have done. But they were things that put us on the map when we needed to be put on the map."

As Smith settled into his new post, a stronger national economy brightened the financial outlook of The Odegard School. And by a stroke of Tom Clifford's famous luck, even the Aerospace Foundation had dug itself out of its huge deficit.

As the contractor of the Spectrum programs, the foundation had used university-owned planes and university personnel in the *ab initio* training. But in 1998, as the China Airlines program was fading away, the foundation didn't have the expected $2.7 million on hand to reimburse the university for its services. Foundation debt had been serious enough that the position Bob Reis held as executive director of the Aerospace Foundation was eliminated at the end of 1997. Later, in early 1998, when the China Airlines program was all but dead, Clifford briefly considered shutting down the foundation. But a decision he'd made earlier in the decade paid off at exactly the right moment.

While he was president of UND, he adhered to a policy of buying up land adjacent to the university whenever it became available. "I always felt you should buy your peripheral land because you're going to expand," he says. "We did, and it saved a ton of money."

In those days, there was a large wheat field across from UND Aerospace on the west side of Interstate-29. It was owned by the Burlington Northern railroad. At the time, Clifford served as a member of the Bush Foundation board of directors in Minneapolis—as did a senior vice president of Burlington Northern. He told

Clifford the railroad's long range plans no longer included the land west of I-29. Did Clifford want it? Indeed Clifford did, figuring it would come in handy someday for the aerospace school. Thus the Aerospace Foundation bought 110 acres of empty wheat field for $330,000.

Toward June of 1997, just after the devastating flood that had closed UND two months early, the university found itself short of cash and was having a hard time making its payroll. At the same time, the city of Grand Forks needed land on which to build affordable housing for the scores of citizens who'd lost their homes in the flood.

Clifford directed the Aerospace Foundation to sell 65 of its wheat field acres to the city for $3.2 million. That cash paid off the foundation's debt to the university, with payment coming on the very last day of the fiscal year. It saved the foundation from what would have been an embarrassing, publicly known debt.

That same year, an unusual benefactor named James Ray came out of the sky, literally, to befriend the school and the foundation and help get them onto a more solid financial footing.

Ray had piloted B-17s during World War II, surviving 30 bombing missions with the celebrated Eighth Air Force. After the war he'd done extremely well in cattle ranching and investing. He'd became a member of the board of directors of Cirrus Design Corporation, the world's second largest manufacturer of single-engine, piston-powered aircraft.

Headquartered in Duluth, Minnesota, Cirrus also operates a large wing-fabrication facility in Grand Forks. When the company held a board of directors meeting in Grand Forks in 1997, Ray made his first ever visit to the city. He flew into Grand Forks airport in his private jet and was immediately surprised by the large fleet of small airplanes taking off, landing and otherwise occupying huge amounts of space at the adjacent UND complex.

"Jim saw the school and said, 'What is this?'" says Bruce Smith.

John Odegard, meanwhile, had already arranged for Cirrus to hold its meeting in his own board room at the aerospace campus. And that is where Odegard and James Ray first met. John, the ultimate salesman gave Ray the grand tour. The flying cattleman quickly fell in love with the school, impressed with the academics, the flight operations and even more so, he said, with the quality of the students.

So taken was he by John and Diane that when Odegard died a year later, Ray made a special contribution to the school. It was Kim Kenville, Tom's sister, working as a fund raiser for the Odegard School in the UND Alumni Office who got the first word of Ray's generosity. In the mail one morning, she found a letter from Ray containing a check for $1 million to endow an Odegard Memorial Scholarship. Ray later added another $500,000 to the endowment. In its first several years, the scholarship covered $80,000 for tuition, books, room and board, and flying fees for a selected student for all four years. Since then, the scholarship has been subdivided and provides significant help to several students from all departments in the Odegard School, not just aviation.

(Kim Kenville later went on to join the faculty of the department of aviation and earned her Ph.D. She is now an associate professor.)

When Odegard died, the close friendship between Ray and the school was fostered by Tom Kenville. Ray's close involvement with the school has continued and since that day in 1997, he has become the University of North Dakota's largest private benefactor. With Ray's continued financial support, the Aerospace Foundation has invested in high-technology training devices and aircraft that now set the Odegard School apart from other collegiate aerospace programs. Ray has endowed students, but he also created a faculty development endowment in honor of Don Smith. His gifts allowed the foundation to purchase Air Traffic Control tower simulators, an air transport aircraft that will lead to the addition of the Eclipse Jets to the aviation fleet, and aircraft for the foundation's extension sites. He had even supported the construction of a hotel connected by the "gerbil tube" walkway to the campus buildings of the Odegard School.

"Jim just felt we needed a hotel connected to the school like the one detailed in John's original concept," says Smith. "He is a close friend of Baron Hilton. Baron goes to the Oshkosh EAA air show every year, where they have a Hilton Garden Inn. Jim was there with him in 2001, and Tom Kenville told him we've got to get one of these in Grand Forks."

Hilton, however, had already passed on Grand Forks, thinking it couldn't support another hotel. That was before the city built its Alerus center, a multi-purpose convention center where UND plays its football games indoors and national bands stage their concerts. Then, the wealthy UND alum Ralph Englestad built a $100 million hockey arena for the university team. Suddenly on sports weekends, hotel rooms were scarce.

It still didn't convince Hilton. So Smith and Kenville went to the Hilton franchise representatives who'd built the Garden Inn at Oshkosh and discussed the idea of the Odegard School building a Hilton on its own.

Smith remembers the representatives saying, "The only way you could do it is if you've got some rich benefactor." Kenville's casual reply, "He's right outside."

Ray was then ushered in, listened to the plan and made a proposal. He would put up a fourth of the $10 million construction cost if the Odegard School would help secure another fourth of the cost from private investors. The remaining $5 million would come from commercial financing. They divided the $5 million from Ray and the potential investors into 100 shares, each one selling for about $55,000. Meanwhile, Ray put his $2.5 million in escrow where it would stay until the entire 100 shares had been covered.

As it turned out, Tom Kenville worked with Ray and sold all but 18 of those shares. The developer offered to buy half of the outstanding shares if the Aerospace Foundation bought the other nine.

"That was a stretch for us," says Smith. "It was almost half a million dollars."

Would the fiscal conservative do it?

"We did it," he says, "It was a huge cash outlay—we paid in increments over five months—and we were right on the edge. There were some scary moments in terms of our cash flow and making payroll. But we made the payments."

A quality hotel adjacent to the school had been a joint dream of John Odegard and Tom Clifford for years. When it was built in 2002, it connected to nearby Ryan Hall by another so-called "gerbil tube." Today, a university guest at the hotel can now access the entire Odegard school campus without going outside. To date, the hotel has filled up consistently, not only for sports weekends but for those many conferences, conventions and concerts now held at the Alerus center, just a few blocks away.

At the moment, the Aerospace Foundation owns a small piece of the hotel, with the possibility that it could divest or even end up a majority owner down the line. It's an option Smith is happy to have.

He scored another revenue win in 2002 that upgraded the school's helicopter program. It started when Smith got a call from Sen. Byron Dorgan's chief of staff in Washington. The senator, long a supporter of UND's Air Battle Captain helicopter training for the Army, had managed to bump up the Congressional budget earmark for the program to $2 million. Dorgan's aide remarked dryly: "You better damn well use it."

Having taught at the Air Force Academy, Smith was aware that military academies conduct special programs each summer for their cadets. At the United States Military Academy at West Point, for instance, cadets planning on going into Army aviation receive no flight training in either airplanes or helicopters. Their first chopper experience comes after graduation when they enter the Army helicopter training school in Fort Rucker, Alabama—where so many UND ROTC cadets excel because of their Air Battle Captain training. Smith decided to fill that West Point gap. He and the aviation department put together a proposal for a 4-week summer helicopter course for West Point cadets, including a chance for each student to solo.

At about that time, Lt. Gen. Jerry Sinn, the comptroller of the Army, called Smith to ask for ideas on what UND might do to train the National Guard and Army reserve in helicopters. Sinn was from Minot and had a son in UND's Air Battle Captain ROTC program. In their conversation, Smith told Sinn of his West Point idea. Sinn just happened to be good friends with the superintendent of The Point and arranged an interview.

The superintendent loved the idea. His staff visited Grand Forks and inspected the program and its spotless safety record. Everything was good to go until Byron Dorgan began feeling the heat from another part of the Army.

"I had 4-star generals in my office over this issue," Dorgan recalls. "Fort Rucker was leery of someone else doing its training. Members of Congress from their area also didn't want to diminish their training. I told them our intention wasn't to injure Rucker. We wanted to advance our abilities and connect with West Point cadets. They would do the pre-training at UND and still go to Rucker. What we did was say, 'Let's try it for a summer.'"

That first summer, West Point sent 25 cadets—the first time in the history of the vaunted military academy that they'd sent cadets to a non-military training program. The cadets liked the UND program so much they predicted that when word got out, UND would become the hottest summer program choice at The Point. They were right. The following summer, 400 cadets signed up for one of the 32 slots available. Over the years the number of cadets trained each summer has risen to 40.

"West Point became ecstatic with the success of it," says Dorgan. "Once you reach a critical mass, it has its own momentum."

Not only was it popular in the summer, but when those cadets went down to Fort Rucker, says Smith, they finished at or near the top of their class—challenged mostly by the increasing number of UND ROTC cadets.

"We can train them much cheaper than the Army can," says Smith. "When the cadets leave here, they skip the basic phase at Fort Rucker and go right into the larger helicopters and begin warfighting skills training. That's where the savings are for the Army. Plus the Army gets top students with a lot of confidence—because they have already soloed in a helicopter. We knew that if we told those cadets, 'At the end of this training you're going to solo in this aircraft,' they would pay attention."

Smith and the North Dakota Congressional delegation combined for another important victory with the military that not only helped the Odegard School, but saved hundreds of jobs for the residents of Grand Forks.

In 2005, the Pentagon's Base Realignment Commission (BRAC) announced that it was considering closing or drastically reducing the mission carried out by the Grand Forks Air Force Base. Opened in 1957, the base has in recent years been home to the 319th Air Refueling Wing. It deploys 42 KC-135 "stratotankers" which refuel the long-range bombers of the Air Force and other aircraft for the Navy and Marine Corps. Removing those tankers would transfer some 2,300 military personnel and their families and about 350 civilian jobs. The base estimated its economic impact on Grand Forks at $350 million annually.

A committee soon formed in Grand Forks to oppose the closure. Smith argued that BRAC had not taken The Odegard School into account when assessing the value of the Air Force base. He helped create a strategy that made the university a prime factor in keeping the base open.

The argument worked. Not only was the tanker shift delayed for several years, but a new mission was added to the base, built around the futuristic unmanned aerial systems technology, or UAS. The actual UAS aircraft—once referred to as "drones"—are known officially as unmanned aerial vehicles or UAVs.

"This propelled The Odegard School squarely into the forefront of a new and exciting aerospace technology," says Smith, "and led to the university's designation as a Center of Excellence (COE) for three distinct groups." The FAA assigned UND its COE for Airspace Designation; The U.S. Department of Defense cited it as a COE for northern border security; and the State of North Dakota named it a COE for economic development.

To get those designations, the Odegard School partnered with UND's College of Nursing for the study of Human Factors, the College of Engineering and Mines for work on payloads and sensors, and the university's Center of Innovation for the commercialization and production. In 2006 alone, those COE designations generated over $5 million in cash and in-kind funding to study UAS technology and deployment in Grand Forks.

According to the Air Force, about 70 per cent of the stratotankers still in Grand Forks will redeploy to other sites in 2009. The other tankers are scheduled to depart in 2012. By that time, the UAS mission would hopefully have progressed enough to cover those losses.

In the meantime, Smith worked out an arrangement with Lockheed Martin Maritime Systems and Sensors to gain access to one of the new UAV's and to receive $1 million of in-kind support for UAV studies. Lockheed was on deck to provide specialized training of students in the operation of UAV's, along with technical consulting.

The fleet at the Odegard School has undergone a futuristic revamping as well. Conversion of the fleet from the old gauge-and-dial cockpit to the all-glass panel with software driven digital readouts and instrumentation is progressing in dramatic fashion.

In 2006, the Aerospace Foundation purchased and leased to the Odegard School seven training aircraft with full-glass cockpit technology. This includes the purchase of four Cirrus SR20 glass-panel aircraft to be used for advanced flight training related to instrument ratings and instrument course work for flight instructors.

Since 2003, UND Aerospace has partnered with Cirrus by training more than 2,500 purchasers of their airplanes at the company's main plant in Duluth, MN. The company also manufactures some of its composite wing assemblies at a plant in Grand Forks, where it employees 600.

James Ray, meanwhile, has continued his goodwill toward the school. In 2006, he donated to UND his deposit, introductory price and essentially his place in line—number 17—to buy one of the very first Eclipse 500 jets off the production line. This unique craft, manufactured by Eclipse Aviation Corp., of Albuquerque, NM, uses the latest in jet technology and simplifies cockpit functions in a lightweight and relatively inexpensive jet.

At the same time, the Odegard School began its first sub-orbital, atmospheric science mission with that on-loan DC-8 brought in by George Seielstad. Having the DC-8, by the way, was another solid argument to the BRAC commission for keeping the Grand Forks Air Force base open. Weighing 160,000 pounds empty, the DC-8 can carry 150,000 pounds of fuel and 40,000 pounds of equipment fully loaded. It needs a runway 12,000 feet long—longer than the 7,000 foot runway at the Grand Forks Airport. Thus, the Grand Forks Air Force base, with its suitable runway 10 miles away, affords it a perfect home.

The school has also expanded in a significant support area. For years, through its scientific computer sciences center (SCC), the Odegard School has written its own software for a variety of aviation training functions—from flight operations to air traffic control simulation to tracking maintenance on its fleet, to scheduling and dispatching student training flights.

Efficiency seems a perfect philosophy for an aviation program. From getting the maximum number of airplanes safely into the blue each training day, to igniting the most academic bang for the federal buck, or to wringing the most air miles out of a drop of fuel, efficiency has served as a rallying cry of necessity at UND Aerospace since day one.

While there are dozens of ways to measure the efficiency of a program, one of the most visible at the Odegard School is the sight of the Flying Team's multiple national championship banners that hang from the ceiling in the foyer of Odegard Hall. For the intense, week-long competition that the National Intercollegiate Flying Association (NIFA) stages each year as part of its Safety and Flight Evaluation Conference (SAFECON), is nothing if not a supreme test of aviation efficiency.

In the years since Kent Lovelace coached the UND flying team to its first national championship in 1985, the school has competed 22 times for the title and won 14 of them—more than any other aviation program in the country, including those from stalwarts such as Embry-Riddle University, The Ohio State University, Auburn University and even The U.S. Air Force Academy and the U.S. Naval Academy. The most recent title came in May, 2006 when the UND team, coached by Assistant Professor Jim Higgins, edged out old time rival Embry-Riddle University-Prescott.

The annual competition tests not only accurate flying and navigation skills, but in written exams it measures the aviation knowledge and pre-flight know-how of a flight team.

Among the flight events in the competition is a test of accuracy in landing. It measures the distance between the touch down point of a plane's wheels and a given runway marker. This is done in two phases—with power on and power off. A navigation event requires student pilots to plot and fly a course, predicting to the second when the plane will reach given checkpoints.

The ground events of the competition test a student pilot's knowledge of federal flight regulations, weather, flight planning, computer accuracy, simulator accuracy and aircraft identification. A pre-flight event challenges students to troubleshoot problems with a grounded aircraft that has been set up with a hidden defect.

"The way we practice for the ground events is to take a ton of written tests," says Higgins, who was a member of two national championship teams as a UND student in the early nineties. "During the day, students go to class and take tests, then they show up at Flying Team practices and get to take more tests. But that's our secret, to be very strong in ground events. It's a mental game. The thing is to recruit people with a competitive spirit, those who have the internal makeup not to accept second place."

The UND Flying team practices about 30 hours a week. It meets early every Sunday morning at the airport at a time when there are no student training flights scheduled.

"On Saturday nights, a lot of students are out doing normal college stuff, having a good time," says Higgins. "But my team can't have beers on a Saturday night and fly on Sunday. That's been tough on a lot of college students, but you have to have a different mentality and mindset. The reward comes when they hang up that championship banner. I ask the students what will be more memorable 20 years from now when you visit campus with your kids? Will it be pointing to a bar where you went for beer? Or pointing to a championship banner and saying, 'I was a part of that'? We want people to take the longer view."

After Higgins graduated from UND in 1992, he became a cargo pilot, then flew for Business Express and became a Captain at American Eagle. He returned to his alma mater as a member of the aviation faculty in 2002.

"You'll see former Flying Team members now as heads of airlines, chief pilots, heads of pilot unions," he says. "They gravitate to the upper echelon because of their competitive spirit. They are virtually unstoppable. The fact that we are here in North Dakota, and we're audacious enough to think our Flying Team can win it every time, is almost a reality. From 1985 to 1992, we won the title seven straight years. No one else has ever won back to back championships. Now we've built up a reputation where people expect it."

While those national championships helped build and maintain the prestige of the school over the past 20 years, John Odegard often bragged about a less visible achievement that has placed UND on a very tall pedestal. As part of his endless promotion of the school, Odegard would often tell visitors and key supporters that the aviation department was about to log 100,000 miles of student flight hours in a single year. In 1990, when the school was first on track to surpass 100,000 hours, the recession and enrollment dip thwarted Odegard's hopes. Still, he frequently left the impression that the 100,000 mile mark had been surpassed. In truth, while Odegard was alive, the program came close but never actually topped that nicely rounded number of prestige.

"John, for many, many years, had been telling people we flew 100,000 hours annually," says Don Dubuque, "but we never did. One year I sat down and figured it out. From July 1, 2001 to May 11, 2002 with our enrollments rising, we actually did fly 100,000 flight training hours."

Dubuque immediately sent the news in an e-mail to the flight operations staff. He even included Odegard posthumously in the e-mail routing with the message, "Hey John, we finally did it."

It was an emotionally-charged moment for those opening their e-mail. Al Palmer, director of flight operations, got the idea that they should carry the message more directly to Odegard. He had the essence of Dubuque's message cut into a small grave stone and, with Diane's approval, staged a ceremony at the cemetery. John Bridewell, a flight instructor who holds a Master's of Divinity degree, wrote a prayer and led a small group in a dedication of that new stone on the fifth anniversary of Odegard's death.

The memorial service happened on a homecoming weekend. "It is just part of our culture to do that," says Bruce Smith.

On Smith's arrival in 2000, the enrollment at the Odegard School was still about where it was in the early nineties, at 1,500. Seven years later, it had grown to 2,100. In 2007, with its fleet now numbering 120 aircraft from North Carolina to Hawaii, the program consistently tops 100,000 flight training hours a year.

The Hawaii program seems to have produced an unexpected dividend. All Nippon Airways of Japan was searching for a school to provide jet pilot training for 200 Japanese college students. Naturally, UND was interested, but when executives of All Nippon came to Grand Forks for a tour, one of their concerns was that their students might feel isolated in such a remote location that seemed lacking in diversity.

But anyone who knew John Odegard, knew of his early and strong belief in a diverse student body. It wasn't so much that he wanted to be a pioneer in equal opportunity, but that he simply wanted to teach people to fly. Anyone who wanted his help, no matter their gender, ethnicity or religion, was welcomed and supported. Through his international *ab initio* programs and global reputation of the school, he attracted men and women to remote Grand Forks from dozens of cultures.

"We don't just teach you how to fly," Al Palmer reminds. "We give you a liberal arts education and teach you how to broaden your mind. We're not just focused on one thing here. You need to know about culture and diversity."

Smith didn't have to make that argument verbally to the visiting All Nippon executives, because a classic Odegardian-style first impression was about to unfold.

At a luncheon for the Japanese executives, they were more than surprised to meet and talk to seven UND students and flight instructors who were from Japan or who had lived in Japan for years and who spoke the language fluently. Several were transfers from that Hawaii community college program.

Thus, in November of 2005, the Aerospace Foundation signed an agreement with Tokai University in Tokyo and All Nippon Airways to train 160 university student pilots over the next four years. The first class of students began their freshman year at Tokai in April 2006. They transferred to UND in April 2007 as full-tuition paying students. They will train and qualify for their flight ratings and, after 14 months, return to Tokai to finish their college degree. Then they will embark on flying careers with All Nippon Airways.

Since 2000, the transition from Odegard's one-man rule to Smith's distributed organizational model has proceeded much less painfully than many once feared.

"The difference between the two of them was just right for the organization," says Paul Lindseth. "It was a young crowd when John was growing the organization. Al Palmer and Don Dubuque and Kent Lovelace and Dana Siewert: they'd been with John for years. Now they're in their early 50s, and they're up for the challenge."

Even the nature of the challenges has changed. For instance, in 2000 when the FAA put out a call to establish a center of excellence in general aviation research, Embry-Riddle approached Smith and asked if UND would like to partner with them on the application. The FAA wanted a geographically diverse group, a partnership between academia, the FAA and industry to solve problems in general aviation.

"Bruce said, 'Let's investigate it,'" says Lindseth. "I don't think John would have done that. He wouldn't allow the formation of a partnership with Embry. We were dreaded rivals. Even I had apprehensions. I probably wouldn't have done it. But Bruce had an open mind that said we're all going down the same path. And it has very much paid off."

The group includes UND and Embry-Riddle, the University of Alaska, Wichita State, and Florida A&M.

UND's aviation faculty has thus far received $3 million in research grants from this collaboration. To date the group's researchers have tackled the safe integration of all-glass cockpit technology into general aviation aircraft, and an automatic direction surveillance beacon to let aircraft in flight see on a flat screen where they are in respect to other nearby craft. Helicopters are already testing the equipment and fixed wing aircraft are next with standardization of methods in place on all aircraft by 2015.

Odegard researchers have also developed new methods of lighting remote airports using reflectors that don't need any power. And they are working on enhanced vision systems that would allow emergency medical helicopters to navigate through low ceiling and low visibility conditions.

"The nice thing is that when you get your faculty energized over these kinds of projects," says Lindseth, "they can hire grad and undergrad students to help them. And those students then become aware of cutting-edge technology."

Answering this type of challenge has enhanced the school's reputation not only worldwide, but even among faculty whose classrooms lie east of the Coulee. The rancor that marked the days of John Odegard has largely dissipated. But while the old misperceptions about low quality academics have faded, some nagging resentments remain—mostly over money.

"That's part of our heritage," says Smith. "In the old days, if we asked the university for something and didn't get it, John's approach was, 'The heck with you, I'll go do it myself.' We're still in that mode. They will say, 'If we tell you you can't have it, you'll just go out and get it yourself.' So your reward for working hard is double edged. Yeah, you have to go get it somewhere else, and that's why you see what we've got here. And that's why everybody is jealous. They say aerospace has all this stuff. But we've had to go get it ourselves. We've always had to do it that way."

Smith feels that many parts of the campus continue to think of The Aerospace Foundation as a private slush fund for the dean. "But the foundation has been steadily growing our assets through the last six years," he notes. "That is reflected in our ATC simulators, aircraft purchases, a new hangar, the Hilton—all are assets of the foundation. And Jim Ray has been a big portion of that in terms of his timely and generous contributions."

Outside contributions, Smith feels, are the direction in which the school needs to go for future viability. Which is why The Odegard School launched its ambitious Omega Fund campaign. It's an international fund-raising campaign seeking to raise $120 million to make sure the school can continue to grow and not have to depend on arbitrary state or university funding formulas to remain in the forefront of aerospace education in coming years.

While there's much going on at the school these days that John Odegard might not recognize or even like, there remains a critical commonality of purpose between his goals and those of Bruce Smith

"With both John and Bruce, there is a central desire to produce a good product for their students," says Dr. Warren Jensen.

Jim Bunke, recently named to the board of directors of the UND Aerospace Foundation, puts it more succinctly.

"There was a real kind of emotional relief to get Bruce there," he says. "We needed somebody, a bright capable leader. Bruce is much more a delegator. But almost all of John's people are still there. They have more responsibility and Bruce lets 'em run. The truth is, years after John's death, the place is still running off of his energy."

EPILOGUE
ODEGARD TO GO

The shades of night were falling east,

As through an alpine village passed

A youth, who bore, 'mid snow and ice,

A banner with the strange device,

Excelsior!

Henry Wadsworth Longfellow
Excelsior

Prior page photo: Even while relaxing at his family cabin on Pike Bay, John Odegard took and made dozens of phone calls. He had one of the first portable phones in North Dakota.

Go confidently in the
direction of your dreams.
Live the life you have imagined.
Henry David Thoreau

Epilogue
Odegard to Go

So, what *would* Odie do? Ten years after the death of the founder, that question still gets asked. In the face of any Odegardian-sized challenge, the lingering spirit of JDO remains the default source of inspiration in the school that bears his name.

For some of his acolytes, the typical Odegard approach to a problem was a simple matter.

"John was never very patient," says Jim Bunke. "He always reminded me of that old cartoon of the two buzzards sitting on the telephone pole. One says to the other, 'Patience, hell; let's kill something.' That was John. 'Let's get the job done.' He approached his battle with cancer the way he approached everything else. He called it the atom bomb approach. He was going to nuke that disease."

In truth, Odegard did have a few more conventional arrows in his quiver. Earl Strinden, for example, often saw a more strategic John Odegard.

"Here's the way John would think," says Strinden. "I was in Bismarck on legislative work one day in the mid-seventies, and I get a call from John. He was in London, about to fly back to the United States on the Concorde. He asked if I would meet him in Washington and set up a meeting with Tom Kleppe, the Secretary of the Interior.

"Kleppe was the first North Dakotan ever named to the president's cabinet. I'd helped Tom in various political campaigns, so I called Tom, and he said he'd love to have us come by to visit. I met John in Washington, and we went over to the Interior to Tom Kleppe's office. He was waiting for us in a big room with a fire going in the fireplace. We sat on sofas and a chair and we visited.

"Then John says, 'Mr. Secretary I'd like to make a phone call.' Tom says, 'Oh sure.' What John did was call an individual in the Interior Department who had major responsibility over research grants pertaining to something John wanted to do. The way it went, John calls this person and lets him know we're up visiting with Kleppe, and John would like to come by and see him. This is the way John Odegard thought. My only purpose in going to Washington was to get John into a chat with Kleppe so John could make a phone call from Kleppe's office, so it sounded like he was a guy who knew the Secretary of the Interior."

That's pretty much the same John Odegard that the pioneering female aviator Jean Haley Harper came to remember. The intricacies of Odegard's operating style didn't really dawn on her, she says, until late in her college career at UND. One day in the early seventies, while she was a student and part-time flight instructor, she tried to renew her instructor's license. To her dismay, she realized it had expired a month earlier. The FAA notified the school that Haley was in violation. She was fined $100, and the school was fined an additional $500.

A hundred bucks was an enormous amount of money to a financially-strapped college student. Devastated, Haley sought out Odegard for advice, but he was away on a trip.

"Weeks later he comes back, all smiles," she remembers. "He told me it was no big deal about the mix-up. He asked me what I had to pay. I said, '$100.' He said, 'What? Why didn't you talk 'em out of it?' I said, 'You can do that?' He told me he sat down with the guy from the FAA and said he was on a shoestring budget, and couldn't afford to pay $500. He gave him a big story, and he talked him down to a fine of $50.

"He could talk anyone into anything, and I soon began to realize he was a deal cutter. He said, 'Next time, argue a little.'"

Former North Dakota governor Allen Olson finds the words 'deal maker' slightly lacking and offers a more dubious description of the problem-solving technique of his old friend. Olson spent four years in the Army's Judge Advocate General's corps as a young lawyer. He came to appreciate a term in use among naval officers to describe a savvy military aide possessed of cunning and street smarts.

"They're called dog robbers," says Olson. "They take care of the admiral. Sort of no questions asked. What they do may not fit perfectly within the regulations, maybe what they do is a little outside the course of military expectations. But the dog robber gets the job done. I had a sergeant who was a dog robber. He saved my tail many times. I think Odegard was the quintessential dog robber. John could always get the job done, but you may not want to know exactly how he got there."

Dog robbing or no, many of John Odegard's most successful principles of management and leadership can be emulated by those who study his life. In fact, many of Odegard's methods came from his relentless study of that mass of Self-Help and How To material that causes so many shelves in book stores to sag.

He was one of those rare executives who not only bought and read scores of books on the best ways to manage a business and get the most from a staff, but he also put those ideas into practice. If he wasn't reading a leadership book, he was likely reading up on new technologies and then doing things like getting the first cell phone in town or the first fax machine.

His self-study only reinforced his belief that nothing was impossible. Almost any topic Odegard became interested in sparked a voracious appetite to learn everything about it and to begin using new ideas. Often he would make use of such ideas on the run, even before they were completely digested. Typical of Odegard is the way he became enamored of aerobatic flying as a young man. He read a book on how to do it and immediately jumped into a plane and took off. With the book strapped to a leg, he glanced at it now and then as he successfully performed the tricky maneuvers described therein.

As a man driven to achieve more than his father had, it's not surprising that he learned never to settle for second best. Whether it was from a book or his own gut instinct, Odegard followed the wisdom once imparted by journalist Herbert Bayard Swope, who in 1915 became the first newsman to win a Pulitzer Prize. Swope rejected the idea that success could be reduced to a formula, but he did offer a brief recipe for failure: try to please everybody. That, Odegard did not do.

Instead he lived a life based on the belief that every day held the promise of opportunity and that those opportunities should be seized without hesitation. He stuck to his belief that every cloud had a silver lining, and that even though a businessman was supposed to be conservative, no business ever achieved greatness by conservative thought or action. Not surprising, his favorite opportunities usually involved spending other people's money.

Odegard also seems to have had no trouble putting into practice a strategy that dooms many a heady entrepreneur: He was never afraid to surround himself with people smarter than he. It's just that he always made sure they knew who was the boss. And The Boss was really The Man who made it clear that loyalty was the first commandment, followed closely by the edict never to utter the words "it's not in my job description."

Odegard was ahead of his time when it came to things like holding staff retreats, thinking big, appreciating the practical impact of looking the part and showing off a little flash bang. That he even had a style struck many academics as unfair and beneath the sophistication of their shut-in world. But Odegard's charisma-flexing and Brooks Brothers suits helped him achieve a cherished goal. For it can be argued quite easily that more than any other native son, he helped put North Dakota not just on the map but on the globe.

"The fabric of his leadership remains," says Dr. Warren Jensen. "We're different, a business married to a philosophy. Do you know the economics of your industry? Do you know the legal concerns? Do you understand human factors well enough to solve the problems you're having with the guy you're working with?

Those were the things he believed in."

Long before marketing became a cultural buzzword and then an essential for globalized economic survival, Odegard grasped the importance of promotion—and he was a grand promoter. But take note: for all the ego necessary for him to think and act large, he was no *self*-promoter. Much is made of the vision required by organizational leaders who would aspire to greatness. Odegard's vision of greatness was not of his own personal glory, but of the quality of his school and the value of the education and career preparation afforded his students.

"He talked to me about goals and aspirations," recalls Jean Haley Harper, "and encouraged me in a way that nobody ever had before."

Odegard scrupulously stuck to the royal "we" in sharing the success of his programs. It was never what "I" have done but what "we" have done. And what was done was always done for his students. Sincerely generous with giving credit to others, he always seemed much too excited and busied by the prospect of the next big thing to waste time basking in any glow. Keep in mind it was the flashy Cray super computer that got the spotlight while Odegard danced on after jets and state-of-the-art simulators.

These common sense traits and strategies which Odegard practiced—dressing and looking the part, remembering people's names, showing enthusiasm, nurturing people—can be found at the top of any list of rules on how to succeed in business. From the favorite habits of successful leaders to the simple philosophies of the "One Minute Manager," the rational guidelines for doing well are no great secret.

But there's a level of achievement deep into the stratosphere above doing well, a supreme level of objective quality that John Odegard reached as an entrepreneur-without-portfolio and which the school that survives him has maintained. The leadership requisites for ascending to this rarefied atmosphere are not simple maxims to be learned from a book. They are inherent characteristics of personality that represent the ultimate opportunistic challenge a human faces: to develop or not to develop raw ability into useful and usable talent.

Start then with the idea of fearlessness. Here is a quality quite different from courage, for one can still feel fear yet perform courageous acts. While the last three years of John Odegard's life were a study in courage and in stoically practicing the positive determination he always preached, it's those fearless years he spent before becoming ill that intrigue.

It isn't that Odegard was simply fearless of the physical danger that all pilots recognize; he seemed to be completely impervious to the notion of getting into trouble—a state of anxiety that most people dread, but which he skipped into and out of time and again with barely a shrug.

You can advise young managers to be willing to try new things and to take calculated risks, but if there is a hesitation gene in their makeup, they won't become John Odegard. Was it his startling recovery from a near fatal episode of polio as a young boy that heightened John Odegard's sense of fearlessness and his unassailable self-confidence? Surely, if one is not born with a sense of fearlessness, or if it isn't burned into the psyche at an early age, it would seem to be fairly impossible to duplicate. Yet Odegard's ability to take physical, social and professional risks calculated to the nth degree—from flash frying a classmate in high school without killing him, to starting an aviation program without any planes, even to gambling that a phone call from the Secretary of the Interior's office wouldn't backfire— suggests the kind of self-confidence that makes one believe he can fly, with or without wings.

"It was as if he saw the world through the eyes of a child," says Odegard's son, John Jr. "He was fascinated by everything around him." To Don Dubuque, his mentor was a dreamer, "but he dreamed in color," he says, which is what excited the people around him to buy into his plan. And he always had plans, this man with a well-known, far-away look in his eyes.

"He would take us out in his boat, and that was his entertainment," says Lyle Samuelson, Odegard's brother-in-law. "But he always had this look. You knew he was busy in his mind."

The busy-ness, the vision, were often grounded in practicality.

"John had the vision to be diversified," says Dana Siewert. "This isn't a pilot factory where we get you a job flying airplanes. He knew there was a lot more to aviation than just sitting in a cockpit. Aviation wasn't a vocation to John. He didn't look at it as a Vo-Tech kind of thing. It was a career opportunity. If all you want is to learn to fly, this isn't the place for you."

To have such dreams is one thing. To follow through on them, as was the Odegard way, took enormous energy—from the boss as well as those who surrounded him. He was a man who, from his wife Diane's testimony, never was able to get comfortable. He simply had to keep moving, and when he did he always seemed breathless with energy.

One of the favorite lessons he imparted to the school is today memorialized in an annual "Above and Beyond" award, given in the shape of a fishing lure constructed of nuts and bolts. It celebrates that day in Snow Lake when Odegard realized he'd brought along his tool box rather than his prized fishing lures. Clandestinely, Jim Bunke and Bob Buley got together and fabricated a fishing lure out of the various nuts and fasteners in the tool kit. Legend has it that in order to ward off the embarrassment, John went out and actually caught a fish using the new lure. The award is given to the person who accomplishes something above and beyond the call of duty.

Every year since 2000, a group of alumni and friends of The Odegard School travel to Oak Lake, Ontario as a memorial to John. The trip has evolved into a scholarship fundraiser and a vehicle to promote The Odegard School to industry leaders.

John Odegard built a strong relationship over the years with the 119th Fighter Wing of the North Dakota Air National Guard, headquartered at Hector Field in Fargo. During a visit in the early 90s Odegard couldn't pass up a chance for a ride in one of the wing's fighters with guard member Jim Peterson. The unit's nickname "Happy Hooligan's" is visible on the tail.

Inevitably, while sitting around a campfire, stories are told of the way it used to be.

"Stuff would happen on the fly," say his colleagues. Odegard, they say, would never take no for an answer, and you could never pry the phone out of his hand because he always wanted to make just one more call. He was an absolutely fierce competitor, he worked harder than anyone in the office, and they all worked hard in that office.

How hard did he work? His close friend Del Rae Meier offers a knowing, bemused assessment. "I don't know anyone who would leave this world wishing he had worked longer," she says, "but I think John was going that way. He loved his work." And that may be the right segue into Al Olson's memory of the man.

"You'd have to work very hard," says this wry, former governor, "to not like John Odegard."

But maybe the better epilogue to John Odegard's career, and to the impact of his dream, is to point to a little known moment of pride that occurred during that horrible morning of September 11, 2001.

On that dreadful day when terrorists attacked the World Trade Center in New York City and the Pentagon in Washington, D.C., the nation's military scrambled to get fighter planes into the air to cover the White House and Congress from attack. Within minutes of the alert, three F-16A jets from the 119th Fighter Wing's "alert detachment" at nearby Langley Air Force Base, jumped into the air, the very first defenders aloft to ride shotgun over the nation's capital.

The 119th is none other than the "Happy Hooligans" of the North Dakota Air National Guard, whose home base is at Hector Field in Fargo. The three jets that raced into the sky that morning were piloted by Lt. Col. Brad Derrig, Major Dean Eckman and Captain Craig Borgstrom. All three pilots are graduates of The Odegard School of Aerospace Sciences.

Derrig, a 1989 UND grad who is now the Operations Commander of the wing, remembers that none of his pilots was fully aware of what had happened that morning to trigger their alert. Once in the air, however, Derrig could see the dark plume of smoke rising in the distance from the Pentagon.

"You get so busy doing what you have to do that you don't have time to figure out what's going on," he says. "We knew something definitely had happened. Something that more than likely would change the world forever."

Not long afterward, Vice President Dick Cheney gave a speech recognizing that the first defenders in the air that morning were from North Dakota. It seemed a bit ironic, given that ten years earlier as Secretary of Defense, in trying to cut UND's Air Battle Captain funding, Cheney seemed to be tarring the whole state with his brush-off comment that the program had "absolutely nothing to do with the safety and security of the United States."

In fact, the school has long embraced and maintained a strong supporting role for the United States military.

"It's a good school," says Derrig. "We'd just gotten the new Piper Arrows when I was there. Everything was very professional. They gave me a strong foundation for military flight school. You could just fall back on that good training."

He says that the actual mission on 9-11 turned out to be mostly routine. Later, he notes, "People we talked to on the ground said they felt a sense of relief in knowing that our fighters were overhead protecting them. I think there was a lot of pride felt by all three of us, being from North Dakota and doing what we had been trained to do all of these years."

Col. Derrig did not know John Odegard well during his UND days, but was introduced to him on several occasions by a friend and fellow member of the Class of '89, John Odegard Jr. A few years back, John Junior gave a talk at a UND Aerospace function in which he spoke of the principles his father set out to instill in him and people like Brad Derrig.

"The key for all of us is not in understanding these principles," said Odegard, Jr. "For the beauty in them is their simplicity. It is not in acquiring them. We all possess them in varying degrees. The challenge is to engrain them into our everyday lives. To really grasp them, add to them the crucial knowledge we have gained at school, both in and out of the classroom, and live them in everything we do. I believe my father can continue to teach this to all of us through the legacy of his character."

AFTERWORD
True Pilots

By Bruce A. Smith, Ph.D.
Dean, John D. Odegard School of Aerospace Sciences

Over the course of nearly 40 years The John D. Odegard School of Aerospace Sciences has evolved from a humble collection of hand-me-down airplanes into one of the largest and finest collegiate aerospace programs in the world. Our visitors are always surprised at what they find here on what they may have once perceived to be barren plains. Almost any measure of success they can name—from cutting-edge facilities at both the airport and academic campus level, to an esteemed faculty, generous and ongoing research awards, state-of-the-art training technology, a continually updated modern fleet of airplanes and helicopters, even to state and federal government-designated centers of excellence—all of it can be found here at the world-class level.

Much more than a pilot-training school, we have built our reputation as a provider of a balanced curriculum founded in the liberal arts. Our flight practice areas are vast, open and uncluttered, all of them adjacent to the quite serviceable Grand Forks airport. We train over flat, open terrain most of which lends itself to the variety of practice landings and takeoffs students need. Our careful training is such that our students seldom find themselves in a forced-landing situation. But if they do, the high probability of landing safely almost anywhere in North Dakota makes our location even more attractive. Even the North Dakota climate serves to our advantage. We may have long, cold winters, but once the snow falls we are blessed with more than our share of sunshine—which equates to a very low percentage of lost flying time over the course of a year. We like to think that the change of the seasons, the challenging weather and the high winds provide our student pilots with a training experience unlike no other.

Earlier in this book, our Flight Operations Director, Al Palmer, attributes the success of The Odegard School to our people. While any organization would likely give credit to its workers, we have made it almost a sacred mission here, from Day One, to view and treat our "workers" as people, and our people as the heart of our work.

Maybe it's an aviation attitude. Certainly the aviation philosophers have noted the special qualities required of those associated with flying. Tom Wolfe famously delineated those who possessed or did not possess "The Right Stuff." In his book "Wind, Sand and Stars," the French aviator, Antoine De Saint Exupery describes the feeling of accomplishment that pilots hold over their "earthbound" comrades.

> "Old Bureaucrat, my comrade, it is not you who are to blame. No
> one ever helped you escape.... Nobody grasped you by the shoulder
> while there was still time to awaken the sleeping musician, the poet,
> the astronomer that possibly inhabited you in the beginning."

And let us not forget the exhilaration that Richard Bach describes in "Jonathan Livingston Seagull."

> "We can find ourselves as creatures of excellence and intelligence
> and skill. We can be free! We can learn to fly!"

Pilots embrace a sense of adventure. There is that special moment of anticipation just before advancing the throttles, and the feeling of exhilaration when the wheels leave the ground. Flying alone, high above the clouds, gives you the same feeling that John Gillespie McGee described as "slipping the surly bonds of earth...and touching the face of God."

There is no feeling of accomplishment in the world like breaking out over the threshold lights at minimums. Once the decision is made to commence takeoff the pilot launches into the virtual unknown. Uncertainty always lurks, but the pilot is reassured by sheer confidence, careful planning, proficiency, and updates from ground agencies. The best pilots show complete confidence in their skills, yet remain conservative in their application. Even with careful planning a potential for change always exists on any flight. Thus, pilots trust the integrity of the people they work with, expecting to receive timely corrections. They expect accurate weather, but seek updates throughout the flight. With their skill and preparation, pilots expect to arrive safely whether at the intended destination, or an alternate or even a road or a field. And if it is an unexpected field landing, that too is okay. A true pilot knows that whatever the decision, it was the right decision and there is no room for being second guessed.

I see the professionals—both fliers and non-fliers—who move The Odegard School forward each day as embracing the possible, much as pilots do. I see them daily exuding the same passion for our mission as a pilot feels for flight. That passion shows in the care our people take in project planning, for instance. Yet countless times I have felt great pride watching them understand that the time has come in the life of a given project to roll the throttles and take off into the bold unknown. Our professors, our flight instructors, our computer experts our accountants, our administrators—every one here has accepted such challenges and, I believe, has felt the same exhilaration as a pilot at the moment of liftoff.

It's the main reason we have won fourteen NIFA National Championships. Or been recognized with excellence awards from the FAA and maintained a stunning safety record. If not for our people it's doubtful we would have heard NASA, China's CAAC, Japan's JCAB, and Norway's AVINOR telling us we are considered one of the best.

Our culture here is one of mutual trust based on a shared professionalism; an individual flying high can count on updates and support from colleagues if conditions change. We've come to view those daily challenges that are part of any endeavor with the sort of creative anticipation as a pilot who realizes the weather is dictating an unexpected change in course.

This is neither accident nor coincidence, but the legacy of John Odegard. *John taught us all to be pilots* whether we flew or not. And so we continue to teach our students to be more than just people who are able to fly airplanes. Our mission—John's mission—hasn't changed: help each student become one of life's true pilots.

Which is why John's memory is still very much alive here in Grand Forks. For his overwhelming personality as the consummate pilot still fills every crack and crevasse of the Odegard School and continues to make a lasting impression on all of the people who knew him and on all who will follow.

ACKNOWLEDGEMENTS

Literally scores of people, from faculty, to administrators, to loyal alums to industry stalwarts helped in putting together this story of John Odegard and the School of Aerospace Sciences. Several deserve special mention.

Diane Odegard gave most generously of her time in helping me chart the high points of John's life and career. Her remarkable candor in recounting intimate stories of family life in the Odegard household freed me to paint a portrait of a man and not a deity. Too often we forget that great men and women are human and face the same daily challenges we all face while negotiating our way to glory.

Tom Clifford's insights into John Odegard's professional style and the mood of the campus during the growing years of the aerospace program were invaluable. Tom, of course, was John Odegard's mentor and enabler. While the state of North Dakota was lucky to have a John Odegard, it was twice blessed with the grace of that nice Irish kid from Langdon.

Bruce Smith, the dean of the Odegard School couldn't have been more cooperative and eager to have John's story told. In his office, Terri Clark's unstinting support stands out as above and beyond—for her honest thoughts on John, a sense of who the key players were and an almost daily font of wisdom and cheerleading. Another supremely busy officer of the school who never failed to stop what he was doing and help was Al Palmer, the director of flight operations. He's also one of the nicer generals you'll ever meet, and I've met plenty.

There are so many others whose kind assistance and cheerfulness have done nothing but confirm in my mind that the good people of North Dakota are one of this country's most valuable treasures. They include Brenda Riskey, Chuck Pineo, Cara Miller, Ken Polovitz, Kent Lovelace, Jane Olson, Don Dubuque, Dana Siewert, Bill Schoen, Jim Kobetsky, Dave Vaaler, Bob Reis, Bryce Streibel, Earl Strinden, Gerry Skogley, George Seielstad, Rodger Copp, Hal Gershman, Lyle Beiswenger, Jerry Nelson, Mark Andrews, Pat Hurley, Robin Silverman, Bob Eelkema, Mike Poellot, Leon Osborne, Tony Grainger, Paul Lindseth, Warren Jensen, Gayle Clifford and Darrol Schroeder.

Even *non-North Dakotans* helped. Mucho thanks beyond the borders of the Flickertail State to Jim Bunke, Jerry Murray, Bob Muhs, Bob Buley, Corky Everson, Jim Buchli, Roger Martin, Russ Watson, Don Smith, Allen Olson, Stephanie Odegard, John Odegard, Jr., Richard H. Johnson, Bridey Orth, Carol Roberts, Steve Larson, Bill Knox, Don Johnston, George and Alma Hammond, Jean Haley Harper, Joe Lapensky, Gary Kiteley, Chuck Kluenker, Robert Shumaker, Kent Alm and Bill Shea.

Patrick A. McGuire
Abingdon, Maryland
June, 2007

INDEX

Page numbers in *italics* indicate photographs.